Russ McElwee

THE COMPLETE GUIDE TO FACTORY-MADE HOUSES

BY THE SAME AUTHOR

How to Avoid the Ten Biggest Home-Buying Traps
The Homeowner's Survival Kit
Dollars and Sense
Buying Land: How to Profit from the Last Great Land Boom
The New Complete Book of Home Remodeling, Improvement, & Repair
Building and Buying the High-Quality House at Lowest Cost

THE COMPLETE GUIDE TO FACTORY-MADE HOUSES

A. M. WATKINS

E. P. DUTTON NEW YORK

For information contact:
E. P. Dutton, 2 Park Avenue,
New York, N.Y. 10016

Library of Congress Cataloging in Publication Data
Watkins, Arthur Martin The complete guide to factory-made houses.
1. Prefabricated houses. I. Title. TH4819.P7W37
693.9'7 79-19580

ISBN: 0-525-93073-6 (cl)
ISBN: 0-525-93085-X (pa)

Published simultaneously in Canada by Clarke, Irwin & Company Limited, Toronto and Vancouver
Designed by Mary Beth Bosco
10 9 8 7 6 5 4 3 2 1
First Edition

CONTENTS

PREFACE: THE KING IS DEAD. LONG LIVE THE KING.

For more than 250 years, the king of U.S. housing has been the friendly neighborhood local builder and contractor who has put up nearly all new single-family houses in the country. His product is called the stickbuilt house. In the past virtually his only competitor was the do-it-yourself American pioneer who put up his own houses, from log cabins to Cape Cod bungalows.

But by the 1970s, houses made whole or in large part in factories were accounting for more than half of all new single-family houses built and sold in the United States. But that's not all. Prefabricated parts and components were being used in most of all other new houses being built each year by local builders and developers. Don Spear, publisher of *The Redbook of Housing Manufacturers*, says that "the term 'factory-made' [houses] could be used to describe up to 84 percent of all new residential construction."

In short, the manufactured house has arrived. Long live the King!

All right, the stickbuilt house may not be totally dead. A small, diminishing number of hammer-and-saw local builders still turn out houses by hand in the same old slow, expensive old-fashioned way, though their handmade houses are by no means as well crafted as they would have us believe.

There is simply no question that a good modern house can be made faster, better, and at lower cost, under the weather-protecting roof of a modern house factory. That is the only logical way to make a house. Continuing to build houses, stick by stick, by hand, one at a time at each building site, makes no more sense today than calling on the horse and buggy to lick a big city's rush-hour traffic snarls.

But "Oh!" so many people cry, prefabs (the widely used though obsolete word for factory houses) are shoddy and poorly built. That's a myth that should once and for all be put to rest. It was started years ago when *some* of the first prefab houses, rushed into production to meet a dire need during and after World War II, were not well made (but then a lot of stickbuilt houses weren't so well made, either). All prefabs were stamped no good, and have since had a hard time living down that stigma. Yet in the years afterward, the quality of design and construction of factory houses has risen considerably above that of locally built conventional housing. Any skeptic who disbelieves this can find chapter and verse on why factory houses are well made in the following pages.

The purpose of this book is to give a fairly fast, comprehensive, and nontechnical look at manufactured houses: the different kinds available, whether you wish to build your own from a factory kit, or from a package, or obtain one completed and ready to move in; the advantages and possible drawbacks of each member of the breed; the savings possible; why a factory-made house is better designed than the typical stickbuilt house; the pitfalls to avoid with some manufactured houses; and, finally, a list of the names and addresses of more than 200 of the leading makers of factory houses.

In all, here is virtually all that a home buyer should know about buying a manufactured house and getting a good one.

PART 1

1.

A BETTER MOUSETRAP: THE NEW FACTORY-MADE HOUSE

After months of frustrating shopping for a good house, a young couple in Maryland—I'll call them Bob and Ginny Simmons—threw up their hands at the idea of buying a house from a local home builder. They had looked at speculative houses and also investigated building a custom house.

Instead they ordered a manufactured house. In other words, a house made by a manufacturer in a factory. The couple found that its cost was roughly 10 percent less than the average builder's house.

"But that's only part of the story," Bob Simmons said. "The factory house gives us more house for our money, its construction details are better [Bob is an engineer and should know] and it offered much faster delivery."

Lower cost, higher quality, and faster delivery are just three of the advantages of a manufactured house, when it is compared with buying a conventional new house made by a local builder—known as a stickbuilt house. Excluding the cost of the land,* a factory-made house can sometimes save you considerably more than 10 percent of the cost of a conventional house. More on savings in a moment. These houses are better built because of the assembly-line quality control possible under a roof and the use of top-grade lumber. And factory

* All dollar figures for houses used in this book will be the building cost exclusive of the cost of the land unless otherwise stated. These costs are those in 1979 when this book was written.

The handsome, unadorned looks of this manufactured house shows its good breeding. Kingsberry Homes, Boise-Cascade Corp.

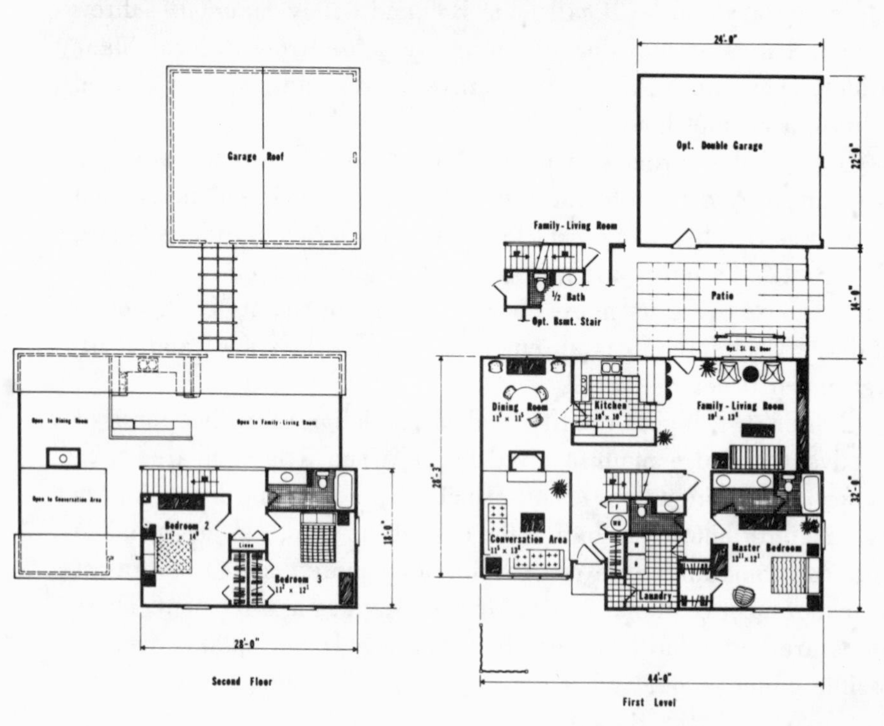

The broad variety of manufactured houses is indicated in this and following photographs. Above is excellent contemporary design (though to keep down heating bills in the North, this house should be located with its large window areas facing the warm southern sun.) New England Components/Techbuilt.

houses can be delivered and made ready to move in faster than conventionally built houses.

Up until the early 1960s houses made either all or in part in factories accounted for a mere trickle of all new single-family houses made in the United States. Now they are streaming out of factories like a flood tide. Some are sold as vacation houses, but the great majority are year-round houses.

In 1978, 612,000 manufactured houses of all kinds were made in U.S. factories, according to John R. Kupferer, executive vice-president of the National Association of Home Manufacturers. That includes mobile, modular, and all other factory-made houses. But that figure is low, says Don Carlson, editor and publisher of *Automation in Housing*, the trade magazine of the manufactured housing business. Carlson says that according to his bi-monthly studies, in

1978, 731,000 manufactured houses of all kinds were made in U.S. factories.

For professionals and others who want specifics, here is a breakdown from these two chief sources of the number of factory houses made in 1978:

	National Association of Home Manufacturers	*Automation in Housing*
Modular houses made in 1978	76,000	141,000
Panelized, precut, all other "prefabricated" houses	261,000	315,000
Mobile homes	275,000	275,000
Totals	612,000	731,000

Total new houses, excluding high-rise apartments: 1,871,000.

No matter whose figures you take, manufactured houses of all kinds now account for a sizable slice of all new single-family houses being built and sold in the United States today. In short, the factory house has clearly come of age.

By 1978 factory houses were being turned out by more than 1,000 different manufacturers—again there's no agreement on the exact number. Another 2,000 or so companies were mass-producing components for houses, such as floor, wall and roof sections, prefabricated plumbing assemblies and complete mechanical cores (also called wet cores) with all the main heating, plumbing and electric parts for a house. These components not only embody the cost and price savings that go hand-in-hand with mass production, but they are usually better made and of higher quality than the same components made at the site of stickbuilt houses by time-consuming hand labor.

But that's not all. Many of the new houses built and sold by the nation's large home builders—those who put up anywhere from 250 to more than 10,000 houses a year—are built with many factory-made components, if not largely built in plants or factories. These are owned and operated by large builders to speed up the production of their houses and to cut their costs by using mass-production techniques. Builders are not dumb. Once their annual volume grows large enough, they are virtually compelled to turn to factory production of their houses to keep down costs.

As a result, it's accurate to say that *factory-made houses now account for more than half of all new single-family houses now built and sold in the United States!*

Don Spear, the expert mentioned in the preface, goes even farther by using a broad definition of manufactured housing that includes mobile homes and all new conventionally built (stickbuilt) houses that use factory-made components. It's on this basis that Spear says that ". . . the term 'factory-made' [houses] could be used to describe up to 84 percent of all new residential construction."

Spear may be stretching the definition a bit, but even so, the manufactured house is clearly here. Long live the King!

Wide Range for Different Buyers

There is a factory house today for practically every kind of home buyer. That includes packages, or kits, for the do-it-yourself home buyer who wants to save money by building his own house (though this is by no means the easy job that many people think). You can also pick your factory house model and package and have a local builder assemble it, or, at the other extreme, buy a factory house that comes virtually complete from the factory.

A factory house also can be bought from a local builder or factory dealer who specializes in putting up one or more factory houses at a time. Some sell a specific manufacturer's houses and will, if desired, modify any model for you. Others build on speculation, putting up factory houses, one after the other, and then placing a "For Sale" sign on each. In fact, a growing number of home builders, from the smallest- to largest-volume ones, are switching to factory houses made for them by manufacturers; more on this follows.

The clean attractiveness of contemporary design is combined here with early American Cape Cod style. Northern Homes.

A Wide Range of Prices

At the low end, prices start at less than $10,000 for complete houses, though obviously not big ones. That little money will get a buyer a compact one-bedroom mobile home adequate for a small family. Like most mobile homes, it comes fully furnished, including kitchen appliances, wall-to-wall carpeting, and a mattress and double bed in the rear. All you need to move in is food for the refrigerator and dishes for the kitchen table.

In addition to mobile homes, there are significant other breeds of factory houses. These fall largely in the $25,000 to $150,000 house range, the same bell curve concentration found in the conventional new-house market, though prices are continually being pushed up by inflation. A growing number of higher-priced models are also made for sale at higher prices.

This traditional house incorporates architectural features of colonial design down to the random oak flooring inside. Scholz Homes.

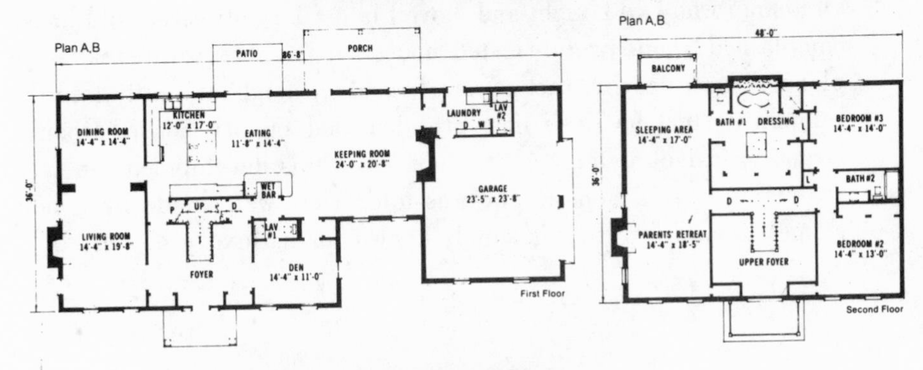

The higher-priced luxury houses include custom-made models. They usually start with a maker's basic plan that is modified in a factory in various ways for different home buyers. Very large and expensive houses are also sold, including models sold for more than $500,000. New England Components/Techbuilt reportedly made the first million-dollar factory house for a Massachusetts family in 1978, though it's not clear how much of that price is accounted for by the cost of the land.

The new factory houses fit naturally into residential America because many are absolutely indistinguishable from conventional site-built homes. Some of their owners, of all people, don't know that they are living in a manufactured house. For example, a well-heeled Ohio man who had bought and moved into a large two-year-old house complained about prefabricated houses, a naughty word, that were planned for a nearby tract of new houses. A neighbor gently told the complainer that he lived in a prefabricated factory house, albeit a large expensive model. "All the houses in this development, in fact, are manufactured houses," he was told. They were made by Scholz Homes, Inc., a firm that is widely known for its luxury houses.

The Acid Test

More and more home builders have not only accepted manufactured houses, but a growing number are using them in developments in the $100,000-and-up house market. Some builders shop for factory houses that are already available and that they like. What's more, manufacturers can not only offer new houses in a wide range of sizes, but they can also offer savings to home buyers. As this book was being written, a Florida development of modest factory-made houses of 1,000 square feet, each one on its own lot, sold for $25,000 apiece. In a nearby development stickbuilt houses of a similar size were selling for $30,000 or more. At a higher level, large factory houses of 2,500 square feet were being sold by builders in southern

California for $70,000 plus land. That was some $25,000 less than the prices of new stickbuilt houses of the same size for sale nearby. Prices for all houses have of course climbed since then.

Other builders bring their plans to a manufacturer who, in turn, makes builder's houses to order in his factory. The builder can have his cake and eat it, too. He puts up the same kind of houses, in design and looks, as he would ordinarily put up himself, and he enjoys the benefits, economies, and speed that are made possible by factory production.

Many manufacturers also make custom-made factory houses for individuals. People bring their plans, including houses designed for them by architects, to manufacturers who will turn out all the pieces and parts for the house. These are then assembled quickly at the site. It's probably only a matter of time before houses made in factories become the widely accepted as well as the predominant kind of housing in the United States.

Perhaps all around the world, too. Foreigners, particularly in the developing countries, are ordering factory houses from U.S. manufacturers (and also from Europe), even though they must pay a stiff extra price of 20 percent or more for overseas shipping charges to their countries. Foreign buyers are continually requesting bids on factory houses from manufacturers. One U.S. manufacturer, Marvin Schuette, president of Wausau Homes, told me that at the time he had some $100 million worth of bids for his houses out for consideration by foreign governments who came to Wausau in need of housing. Manufactured houses have not only passed the acid test of homebuilder approval, but picky architects and skeptical foreigners have also become believers.

The idea of making homes in a factory and shipping them to the point of use is by no means new. The first prefabricated house in the new world was made in England about 1670 and shipped to Cape Ann in Massachusetts. Others followed, including a number sent to early Cape Cod settlers. More than 500 prefabricated houses were shipped to California from New York during the 1849 Gold Rush; more were shipped there from Europe and from China. In the 1890s, at least two U.S. manufacturers were turning out prefabricated houses

The factory production of houses is now a big league industry. This Columbus, Ohio, factory is one of more than 600 in the United States. Inside of the factory a large roof section is being put on modular house unit going down the assembly line. Finished modules are trucked down the highway and stacked in place by a crane at the site. Cardinal Industries.

on a paying basis; one of them, Hodgson Houses, is still in business in New England.

Four Main Kinds of Factory-Made Houses

They are: modular (also called sectional), panelized, precut, and fourth, the mobile home. The first three differ in degree of completion of the house package when it leaves the factory. The modular is the most complete, a three-dimensional house 95 percent complete when it comes off the assembly line. It is shipped in two or more sections for set-down at the site. There it is hooked up to electricity, water, and other utility lines required to light the house and make the appliances work. Then you can move in and turn on switches in the twentieth-century way for things to work while you read the newspaper.

Panelized and precut houses are made in two successively reduced stages of factory completion. Each thus requires respectively more completion work at the site. Fourth is the well-known mobile home, scorned by many observers but lauded by others, including many buyers. New versions of it being introduced are low in cost and

CARDINAL

A growing number of manufactured houses are also being made in other countries and particularly in Canada and England. This is a three-bedroom factory house made with a panel system (also shown) by British manufacturer, Guildway Limited in Surrey, England.

attractive in looks. There are also log cabin, dome, and A-frame houses. The last three categories, however, have to do with the type, style, and architectural design, rather than the method and extent of factory production.

Some manufacturers of factory houses other than mobile homes do not, by the way, accept the mobile home as a "manufactured" house. They acknowledge only three major forms of factory-made houses: modular, panelized, and precut. This relegates the mobile home, by implication, to an inferior category of house. Yet by Webster's definition, a mobile home is not only made in a factory, it is also more completely made there than any other kind of manufactured home. It is 100 percent factory made, a manufactured house by definition. More and more mobile homes are not only attractive, they are being anchored to permanent foundations and are financed and taxed in the same way as conventional houses. There are without question four breeds of manufactured houses and each is dealt with in this book.

More significantly, it is clear that houses are today finally being made logically. Buying anything else but a factory house today is as old-fashioned and obsolete as buying a house without running water.

That's no big surprise. It was predicted by eminent architects, including the famed Frank Lloyd Wright and Walter Gropius, a leader of the celebrated Bauhaus movement in Germany before Hitler and later an eminent professor of architecture at Harvard. Each not only believed in the cause of factory-built houses, but also designed pioneering houses for mass production in house factories. Amen.

2.
SAVING MONEY WITH A FACTORY-MADE HOUSE

A young couple in Washington, D.C., had an architect draw up plans for their dream house, but the lowest bid to build it was $130,000. That was, alas, more than they could afford.

A friend suggested buying a factory house. To their surprise and pleasure, they found some that were as good and attractive as their unborn architect-designed house. They bought a factory house at a price of $105,000, completely built. That saved them $25,000 on the house, not counting several thousand dollars of welcome savings on their reduced architect's fee.

That's an example of the large savings possible with a factory-made house. That's compared with the price of building your own custom house, whether you build from a stock plan, or from your own architect-designed plan, or buy a custom house from a builder. Not every factory house, it is true, may save you that much in actual dollars on its initial cost. How much you save "is a very hard thing to say," says L. Michael Carusone, president of Northern Homes of Hudson Falls, New York, a manufacturer of especially well-designed houses.

The savings vary greatly according to such things as the type and brand of house and where it's sold. Carusone adds that two of the major savings made with factory houses result from their high-quality construction and the speed of completion. These can mean significant money savings for reasons cited in chapter 3.

The actual dollars saved on the purchase price of a factory house may range from nothing at all for some to a great deal, according to different sources. A spokesman for the National Association of Home Manufacturers says that, excluding mobile homes, the savings "average about 5 to 10 percent." Other sources say that they can go up to 30 percent, more or less. The largest savings of all are made with mobile homes where the savings average 46 percent, as shown later.

There is no agreement on the savings you can make because they depend on the turns and bends in the road on your way to buying a particular factory house. There are, unavoidably, a number of variables. Understanding what they are, however, can shed light on factory houses and show you how to save the most when you buy.

Consider first, house quality. Good-quality construction costs more. One manufacturer says that this means reduced savings for this reason: "Take all the 2 x 4 wall studs used in our houses. They're Number One grade, and recently they cost us $1.25 apiece. Local builders in this area, our competition, use utility grade 2 x 4s at a cost as low as 69¢ each. That's a big difference. We pay up to twice as much for good lumber. It really adds up when you consider the hundreds of studs that go into the walls of a house.

"Now take into account the extra cost for the extra quality that goes into the other wood and other materials throughout the house. That's no small potatoes, since some 10 to 15 tons of building materials go into a house. That's a significant extra cost item for the factory house. It offsets some of the savings made as a result of factory houses being made faster and more efficiently than stickbuilt houses."

It may also be true that some good builders also use top-notch Number One lumber in their stickbuilt houses, as is claimed by some. But then their house prices will also be higher than the run-of-the-mill locally-built builder house. Comparing these better-made stickbuilt houses with equivalent factory-made houses is now comparing apples with apples. In this case, the factory-made house will almost always be the low-cost winner, since it can be made faster, more efficiently, and at lower overall cost in the factory.

Savings will vary according to the degree of completion of the

This is an example of the increasing number of mobile homes that are indistinguishable from conventional houses. Its cathedral ceiling and airy kitchen-dining area are also shown. Reprinted from Better Homes & Gardens.

house when it leaves the factory. Mobile homes and modular houses are the most fully made in the factory and therefore require the least high-cost, on-site labor after delivery. That's a major reason why the mobile home, the most completely made factory house today, offers the greatest savings; modular and sectional houses—about 95 percent complete on delivery—offer the next greatest savings.

You will save progressively less money with a panelized house and least of all, with one exception, when you buy a precut house. The precut is the skinniest factory package, and therefore requires the most on-site completion work. The exception is building your own precut house and thereby saving a lot as your reward for using your own do-it-yourself labor and elbow grease.

Custom Changes

Asking Ford or General Motors to stop its assembly line and make custom changes on the new car you ordered would cost a small

fortune. The cost of your car, stopped, in effect, in mid-air while it's being modified, would not only go up sharply, but the cost of delays with all the other cars on the line would also rise.

In similar fashion, asking a home manufacturer to make changes in one of his standard houses can also cost extra money. These might be things like adding an extra room or just another closet or two. Whatever they are, they mean stop-and-go interruption with the production flow of a house being made. Changes requested might also require special plans and specifications to be prepared first by the architectural department, and then other people in the plant must also change gears to accommodate the changes.

Unlike high-volume, highly sophisticated Detroit assembly lines, many house factories can make custom changes in their houses to suit home buyers. The factory production of houses is still in its relative infancy and most house assembly lines operate at comparatively slow speed. Nonetheless, custom changes in manufactured houses costs extra money, and hence affect savings on the factory houses involved. For that matter, asking a local builder to make custom

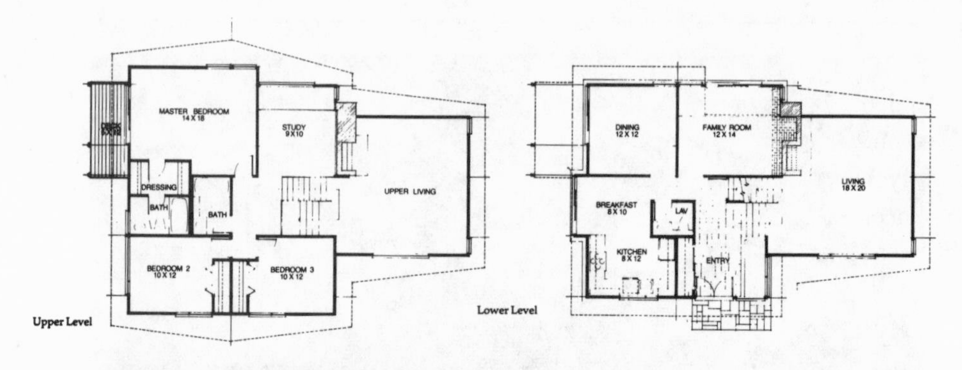

Handsome factory-made houses like this one offer excellent design and other advantages at substantially less cost than a one-at-a-time architect-designed custom house. Deck House.

changes in one of his stickbuilt houses will also mean a higher price for the house. Moral: keep your custom changes to a minimum. Shop for a maker's model that is suitable for you with few or no changes.

Delivery costs can eat into savings. The longer the distance a house must be shipped from plant to site, obviously the higher its shipping cost. In one way or another, this is paid by the buyer. The maximum distance manufacturers generally ship is about 500 miles and many limit their deliveries to shorter distances. Because there is no standard shipping distance, delivery costs vary from buyer to buyer. Thus savings will vary for different buyers on total house price.

Dealer Markup

Markup is another widespread variable which, like a seesaw, affects house prices. The more a builder-dealer marks up his sales price, obviously the less the buyer saves compared with buying a stickbuilt house. Some makers put firm prices on their houses and builder-dealers must toe the line to these prices. The retail prices on other factory houses are, for various reasons, set by the local builder-dealer. As a result, a builder in one area may charge more for a particular house than another builder in another city for the same house.

Sometimes, to put it bluntly, some builder-dealers may jack up their factory house prices as high as the traffic will bear. This, unfortunately, can wipe out a good chunk of the savings otherwise possible on a factory house. The builder could sell for less, perhaps. But why do it, he asks, when local stickbuilt houses sell for more and he can get just as much for his houses? This age-old practice of human take-all-you-can-get is common, of course, nearly everywhere merchandise is bought and sold. In the future, however, it may be largely eliminated among factory houses as more and more factory houses hit the market and competition among them increases. Your best defense now, like getting a good price on other merchandise, is to shop around for the greatest savings.

Finally, savings on factory houses can vary according to local construction costs and prices for stickbuilt houses. The higher the cost to build a stickbuilt house, naturally the higher its sales price will be. And local construction costs fluctuate considerably from one place to another. Factory house prices are more uniform nationally. Thus, you can save more on a factory house in an area of high-cost stickbuilt houses.

In addition, Douglas Scholz, president of Unibilt Industries, Inc., Vandalia, Ohio, says that getting a good house at an affordable price is one thing in large cities and their suburbs, but often quite another in small towns and rural areas where ". . . there are simply not enough skilled home builders to produce the kind of house people want." A factory house, particularly those that are made and shipped virtually complete, therefore offer especially welcome savings in such areas.

Factory house savings can, on the other hand, diminish in larger cities and dense suburbs where new houses are built and sold by high-volume builders. In effect, they produce their own efficiently made factory houses, thus at reduced sales prices. The only difference, as noted earlier, is that such a builder's plant or "factory" mass produces his own houses only.

Vacation House Savings

A manufactured house is clearly a natural for a vacation home. It's a monarch in this area. At one swoop, it does away with most of the complications of building a house in a distant rural area, e.g., finding a good builder and skilled workmen, to start—not in abundant quantity almost anywhere—and then riding herd on your construction gang by long distance. Most expensive of all, you can easily be a victim of the soak-the-city-slicker syndrome on everything that goes into a vacation house being built stick by stick.

A factory vacation house can clearly save a small bundle of money compared with a stickbuilt house. Because some factory houses especially lend themselves to do-it-yourself completion, greater than usual buyer-builder savings can be made on this score. A factory vacation house often can be completed while you're using it. A number of home manufacturers started in business by turning out vacation houses, and vacation houses from factories are now available nearly everywhere in the country.

Building Your Own Home

Savings also can be made, naturally, by building your own year-round home from a factory package or kit. Building any house is no easy chore. Don't kid yourself about this. It takes work. But it's considerably less of a mountain to tackle than building a stick-built house.

Manufacturers point out that do-it-yourself building savings are largely labor savings, i.e., the result of furnishing your own no-cost labor to finish the house. Often, they say, having professionals do the hard construction things is still recommended. That can include putting in the foundation, erecting the basic structural shell, and installing difficult things like the plumbing, wiring, and heating. Finishing the rest of a house is then easier for a willing though inexperienced buyer.

This has led to the shell house phenomenon. Manufacturers sell factory houses that are completely built on the outside; in other words, the structural walls-and-roof shell is built and closed in for you. Both small and large companies like American Barn/Habitat, Boise Cascade Ridge Homes, and Timberpeg offer such shell packages. (Details about these and other companies cited in the text are given in Part II, the Directory.) The buyer takes it from there, completing the rest of the house himself. Some makers, like the Heritage Home people, are totally flexible. Each buyer may decide on how much of the total construction the factory will provide and how much the buyer provides. Naturally, the buyer saves on what he chooses to do.

The Biggest Savings

You save the most when you buy a mobile home, the lowest-cost house of all. Remember that there are mobile homes and mobile homes. Some of the new ones can take your breath away. They are no more like the old house trailer—the Model T of manufactured houses—than an upper-crust residential road resembles shantytown. An increasing number of mobile homes are, in short, as attractive as a cover girl, and they rank right up there with the most attractive houses of all.

Not all mobile homes are beauties. Many are still boxy, tinny, and grim looking. They may be dolled-up [sic] with flashy flower pots, wrought iron or other warts that are ill advised. But just as clothing can run across a spectrum from the cheap and gaudy to the handsome and attractive, so can mobile homes.

A nice thing is that the really good mobile homes do not necessarily cost much more than those of questionable design. This applies chiefly to the homes, and not necessarily to mobile home parks and developments. Renting a site in a good mobile home park usually costs more than renting one in a run-of-the-mill park. In some areas of the U.S., you also have to shop a little harder and longer to find a good, if not top-quality, mobile home. But it can be worth it.

On average, a mobile home can save as much as 46 percent of the price of a locally built conventional house. In other words, savings run almost half the price of a conventional house. That's based on figures from the U.S. Department of Commerce construction-cost division in Maryland, the acknowledged national oracle for building statistics. In early 1979, its surveys found that mobile homes sold for an average of $16,000, or $15.50 per square foot of usable living area. The average new site-built house sold for $29 per square foot, or almost twice as much. Actual home building prices for stickbuilt houses range from $25 or so per square foot in low-cost construction areas to more than $40 in high-cost areas.

Summed up, a good mobile home is just about the biggest house bargain anywhere. Where else in the world can you buy a furnished three-bedroom house with central air-conditioning for as little as $16,000?

Answer: nowhere else.

Incidentally, nearly all mobile homes are mobile only for their brief initial birth period. That's the time from factory completion to overland delivery to the first buyer. Once bought, lowered onto buyers' sites, the wheels taken off and the unit tied down, most mobile homes are never again moved. In addition, a growing number of mobile homes are being anchored down to permanent foundations, like conventional houses. It would take the force of a bomb to move them again.

In sum, no general statement can be made about how much you can save with a manufactured house. Clearly, it depends on the kind of house bought and the other variables mentioned. Besides the actual dollars saved, other special advantages of a factory house can add up to real money in the bank, major savings that are spelled out in the next chapter.

3.

TEN MONEY-SAVING BENEFITS OF FACTORY HOUSES

This chapter is a paean to the factory-made house. It's a technological accomplishment that was long overdue. The handmade stickbuilt house, by contrast, is as much an anomaly from the dark ages as light from a candle.

Still, the art of efficient mass production of houses in factories is still not perfected. It's in a relatively early stage of its development. Improvements are to come, so don't expect a perfect product now or a rock-bottom price, or both. At this stage of its development, though, the factory house is a huge step forward for home buyers with the following major benefits.

High-quality construction. This ranks well up there in importance with money savings. Top-grade lumber is used in nearly all factory-made houses, largely because second-grade lumber can cause problems with the precision techniques in the plant and thereby slow down production. Most factory houses also are built to conform with the toughest building codes.

Special rigidity and toughness also must be built into factory houses because of the nature of factory-house production, and particularly to withstand the bouncing around encountered during over-the-road shipment. Practically no stickbuilt house could stay together, for example, after being lifted into the air and dropped a few times, but many factory houses can. Such strength is overkill, you might say, since Paul Bunyan is not likely to pick up a house, like a ball, and bounce it around. Right?

Wrong. Such punishment could be dished out by a twister, a landslide, or an earthquake and many a factory house can take such a punch like Muhammad Ali. For example, not long ago a modular house was delivered to its site too late in the day to be anchored down. A tornado came across the country and lifted the house in the air, spinning it around three times before dropping it to the ground. Peering from their storm cellar, the people next door saw it happen. The next day the assembly crew found the house intact except for a few dents and broken windows. The house was hauled back in place—it had landed halfway off the foundation—and was anchored down. The dents were repaired, glass replaced and the house was as good as new. It has been occupied ever since. The maker of the house says that that's his answer to the Wizard of Oz.

Another high-quality feature with welcome results is the kiln-dried lumber used in factory houses. That means predried. One couple I know did not know about this till a year after moving into their new factory house. A relative in the construction business was visiting. He said, "It's amazing. You have no cracks in your walls. When my new house was a year old, there were cracks all over because of the wood drying out."

That doesn't happen to a factory house because the predried wood does not shrink and warp as a result of uneven drying. Predried wood also means no squeaky floors and no bulging walls and popping nails.

High-quality design. Don't underestimate this one either. A sad fact of life in homebuilding is that professional architects are hired to design only a small minority—perhaps 15 percent—of all conventionally built houses in the United States, according to the American Institute of Architects (AIA). Alas, the haphazard results show it. Many manufactured houses, on the other hand, are designed by architects, including top-drawer residential architects. The manufactured house comes with better breeding. It shows both in form and in function. That can mean better long-term value, in other words, higher resale value, as well as the day-to-day benefits of living in a better house.

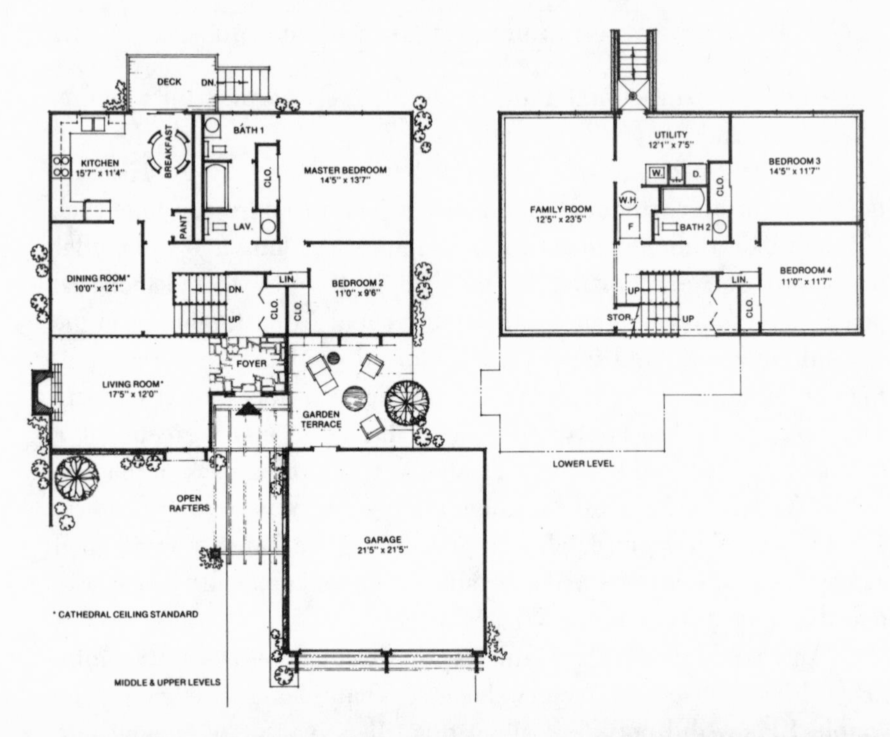
DECK
DN.
BATH 1
BREAKFAST
KITCHEN
15'7" x 11'4"
CLO.
MASTER BEDROOM
14'5" x 13'7"
PANT
LAV.
DINING ROOM*
10'0" x 12'1"
DN.
UP
LIN.
CLO.
CLO.
BEDROOM 2
11'0" x 9'6"
FOYER
LIVING ROOM*
17'5" x 12'0"
GARDEN
TERRACE
OPEN
RAFTERS
GARAGE
21'5" x 21'5"
* CATHEDRAL CEILING STANDARD
MIDDLE & UPPER LEVELS
UTILITY
12'1" x 7'5"
W.
D.
BEDROOM 3
14'5" x 11'7"
CLO.
FAMILY ROOM
12'5" x 23'5"
W.H.
F
BATH 2
BEDROOM 4
11'0" x 11'7"
UP
LIN.
STOR.
UP
CLO.
LOWER LEVEL

Handsome Florida house is built the way houses should be on a beachfront with the ever-present danger of flooding—on stiltlike piles. This can prevent a house from being washed away by a bad storm. Cardinal Industries.

(Facing page)
Proven design is characteristic of most factory houses. This one turns an attractive facade to the public street and is oriented and opened up to the owner's domain in the rear. Pease Company.

A factory house is no pig in a poke. You know what you're getting and thus no painful surprises later. The same house has been built before and the bugs ironed out. You can often see exactly what you will get by seeing the same house on model-house row outside the factory, or by visiting a previous buyer who owns one. At least one maker (U.S. Homes; see the Directory in back of this book) issues a booklet list containing several thousand names and addresses of the buyers of its houses, and it shouldn't be hard to find some close to your home. Seeing an owner can pay off well, since you can find out firsthand how the house has worked out. What should *you* know before you buy? Are there any shortcomings? Any problems to avoid? A past buyer can answer such questions.

Known sales price. The price of a factory house is, unlike putty, usually firm. Few gremlins are likely to torpedo your building costs. Like other consumer products with fixed price tags, it's usually easy to put a firm price on a factory house and stick to it. As many home buyers ruefully say, building or buying a new stickbuilt house often involves paying extra for cost overruns. (These are not always the builder's fault. They're often due to the crazy-quilt nature of stickbuilt construction.)

Building time is sharply reduced. A factory house often can be delivered and ready to occupy within two to three months after ordered. Sometimes it can be as little as a week or three after the foundation has been completed and the house delivered. That's significant compared to the usual four-to-six month completion time for a stickbuilt house. Getting a stickbuilt house finished even in six months can sometimes be cause for rejoicing, as many a wait-weary buyer can attest.

The time saved on a factory house can mean lower interest cost for the construction loan required to build a house. That's the money that a builder or buyer must usually borrow to finance the building of a house, and a higher interest rate is charged for it than for mortgage loans (which buyers use to finance the purchase of the completed house). This can add up to four-figure savings. Building overhead costs are clearly lower when a house is finished faster.

True, these and other construction savings may be builder costs that end up in the builder's pocket. But just as practically every building cost for a new house is paid for in one way or another by the buyer, savings made are also reflected in the bottom line, i.e., they too usually come back to the buyer. And more directly, the buyer saves by moving into the new house faster, thus leaving behind his existing house sooner.

Made-to-order house packages. Some manufacturers will on order turn out a house package built according to a buyer's plans and specifications. They are called *custom* manufacturers. Each will make custom house packages for individual home buyers, including enough houses to stock a new development for a builder.

There are also *catalog* manufacturers who turn out a line of houses shown in their catalogs. You may buy any one, often with optional changes, if desired. Many makers will also modify their standard plans on request, though generally this costs extra money. Some manufacturers do both. They make their own line of houses shown in their catalogs and on request will produce custom house packages for buyers.

Easy financing. Many banks and other mortgage lenders readily approve a mortgage loan on a factory house because the house has been evaluated and financed before. The factory house generally comes with first-class credentials, including certificates of approval by state, regional, and national building codes. The house construction is clearly acceptable. A buyer might also be blessed with favorable terms, and the mortgage application ordinarily sails through processing. Like any mortgage, the pain comes every month later when you must pay it back.

Building your own factory house from a kit is practically child's play. Okay, it's still hard work. But, nonetheless, it's considerably easier than building a stickbuilt house. And, as noted in chapter 2, a home buyer can practically call his own signals and pay the manufacturer only for a half-completed shell house, and complete that or any other part of the construction with his own money-saving labor.

Reduced theft and vandalism during construction. Many factory houses can be closed in and roofed over very quickly, some within a day or two, and therefore locked up quickly. Thus, it's no sitting duck for thieves and vandals, who unfortunately are major home-builder vultures nowadays. It's a one-day closing-up job for Boise-Cascade's 1,200 square-foot houses. (See photographs in chapter 5.) A company official says, "The same-size stickbuilt house requires up to ten days for the same degree of completion." Closing up a new house fast not only cuts down on theft, a builder adds, but work can start inside at once no matter what the weather.

Sharply reduced waste. Building one house at a time, as many builders do, ". . . can lead to very expensive waste and scrap," says T. W. Cahow, president of Continental Homes of New England and a past president of the National Association of Home Manufacturers. He says that leftover wood, nails, and broken cartons of materials like flooring and roofing, "are often carted off as scrap, adding nothing to house value, but certainly adding to the cost."

In addition, lumber and other building materials stored outdoors can deteriorate, if not spoil like food, through exposure to the weather. Waste can be kept to a minimum with a factory house because each step of the construction has been planned in advance. The lumber and other materials are supplied in the measured quantities needed.

Saltbox with deck fits naturally into New England vacation setting. Acorn Structures.

Warmth of wood and open planning make this interior large and appealing. Deck House.

(Above right)
Two-story house fits naturally in its forest setting. American Barn/Habitat.

Most factory houses conform with Federal Housing Administration (FHA) standards for houses. That's in addition to conforming with state and regional building codes. It's another step forward because it makes houses eligible for government-insured mortgages, like FHA, Veterans Administration loans, and Farmer's Home Loans. It's also a significant plus because meeting FHA standards can also mean easy approval of a private mortgage loan from a private bank or other lender. Many private lenders use FHA construction standards as a basis for approval of the mortgage.

Those are eleven advantages or benefits of factory houses—one more than promised. But be realistic. There are also a few bad apples among the manufacturers and lemons among the tens of thousands of new factory houses turned out every month; it's inevitable. But the number of badies is far fewer than those encountered among stickbuilt houses because the nature of factory production sharply closes off the openings for mistakes. Wise shopping and checking before buying are still recommended.

Drawbacks and Limitations

Deep as one may probe, it's hard to find drawbacks with manu-

Prototype house is built with four factory-made modular sections that can be assembled in one day, thus great savings. House was developed by Popular Mechanics *magazine and the Armstrong Cork Company and made by Fuqua Homes, Arlington, Texas.* Armstrong Cork Company.

factured houses. Contrary to public misopinion, most are very well made. Not all, however, may be blue ribbon winners in the good looks department. The worst of all, unfortunately, are some mobile homes that have not progressed much since the Model T days of the breed. Taste, of course, is a personal thing. And there is a wide choice across the spectrum in factory house style and looks.

The floor plans and room layouts of some manufacturers may offer less variety than stickbuilt houses. This is an occupational hazard in the business because the nature of mass production favors standardization. Nonetheless, manufacturers are beginning to break through such limitations and develop more varied floor plans.

Factory houses are still not sold all over, unfortunately, which is a frustrating limitation for many potential buyers. Factory houses are still barred in certain cities and towns by arbitrary building codes and other silly rules which were often written, in effect, by entrenched local unions, contractors, and building-supply power interests. They can't compete with factory houses so they wage dirty war against them. Two notorious places that bar factory houses are Chicago and New York City, both of which are cursed by building codes that are

relics of the dark ages. That's one reason why local taxpayers in these cities pay through the nose for their housing.

High-quality construction is virtually inevitable when structural parts of a house, like these wall panels, are made square on jigs on the assembly line. They are locked into place for virtually flawless joints. Pneumatic hammers shoot nails into each joint with machine-gun speed.

Working indoors in a well-lit, weather-protected environment, worker saws opening in wall sheathing for window. Automatic nailer in background nails down wall sheathing skin to panel, with a score of nails driven each time.

4.
HOW TO TELL THE PLAYERS WITHOUT A SCOREBOARD. FIRST, THE MODULAR HOUSE

Here and in the next three chapters are profiles of the four main kinds of factory-made houses: modular, panelized, precut, and mobile homes. The first three vary chiefly according to the extent each is made in the factory before delivery, and conversely, how much completion work is required after delivery. But once completed after delivery, they all look as if from the same mold. Only an expert can tell their different genealogy. The mobile home is more easily spotted, though a growing number of new mobiles are quite attractive and taken for conventional houses.

Knowing the basic differences among these four kinds of houses is important. Among other things, it can help you save the most money on a new house and get the most suitable house for your needs.

The Modular House, Crown Prince of the Breed

Also called a sectional house, the modular house could well become the supreme king of housing. It's strong and well made, and made virtually completely in the factory. It's made fast and efficiently and, as a result, usually offers good savings to buyers. Like all factory houses, it's set over a standard foundation, with or without a basement. It is financed by the standard amortized home mortgage in the same way that conventional houses have long been financed.

Like most manufactured houses, most modulars conform with state and regional building codes. Actually, it is built stronger and tougher than virtually any other kind of house, as demonstrated by the minimal damage done to a modular house by the tornado mentioned earlier. Superstrength is built into modulars not only so that they can endure possible pounding during long-distance shipping, but also so they can be lifted bodily by cranes. One expert says that if you, or any unfriendly cyclone, tried to lift the average stickbuilt house, the result would be "instant kindling wood."

Quality Control

A modular house is subject to quality control checks by inspectors at stations along the assembly line. Such control over mishaps [sic] is virtually unknown with stickbuilt houses. At the end of the line, a chief inspector, clipboard and checklist in hand, walks through and inspects each item on his list before he can tag the house with a completion slip.

Modular and other factory-made houses, by the way, are also inspected in their factories during construction by independent building inspectors, called third-party inspectors. This is to check on behalf of home buyers that factory-made houses conform with state building codes where they are sold.

The modular house comes off the factory assembly line made in two or more finished sections, usually 12 or 14 feet wide and up to 60 feet or so long. In other words, it's a three-dimensional package. As they come off the assembly line, the sections are loaded on flatbed trucks and head down the highway. Two or more sections are connected together at the site to form a full-fledged house. Additional sections are added to form larger houses, including stacking the sections to make two- or three-story houses, or higher. Once delivered, a modular house can be ready to occupy in a few days. That's lightning speed, of course, compared with the months required to finish a stickbuilt house. Outside utility lines for water, gas, electricity, and sewage disposal must be connected to the house, of course. The main intake lines for the wiring, plumbing, and other utilities inside the house, which were installed in the factory, are capped and so are ready for outside hookups. Each is uncapped after delivery for connection to its outside line.

The delivery time and occupancy date of a modular house obviously depends on making the foundation ready for the house in advance. Time must also be allowed to bring in the outside utilities for the house. Some modulars may require certain interior needs to be provided after delivery. An example is the unfinished modular which is sold for less money with the proviso that the buyer will finish the interior with his own labor.

The foundations for modular houses, like those for other factory houses, must be made accurately aligned and level. This is essential or the house will not fit on it properly. If the foundation doesn't conform, you're in trouble.

The word "modular" has to do with a standard basic dimension for the construction of houses. It goes back to Eli Whitney's principle of the interchangeability of parts which revolutionized factory mass production in the nineteenth century. All the key structural parts of a house conform with a common basic dimension, such as a 4-inch module or a multiple of it. All the different pieces then can fit together in different ways, just as the different pieces of a child's Tinker Toy set, a marvel of modular design, can fit together and form an enormous variety of different little structures.

For a house, a small number of standard modular parts for the floor, walls, and roof can also fit together in many different ways. Using the same modular house parts, therefore, a manufacturer can easily turn out different houses in a great variety of different sizes and shapes from a small number of the same internal bones and framing pieces. To a home buyer, that also means that a manufacturer often can modify one of his standard house models at minimal extra cost.

Like most manufactured houses, modulars can be bought in most parts of the United States. Don Carlson, editor and publisher of *Automation in Housing*, points out that all conform with regional and state building codes. As noted above, independent inspectors check them in the factory while they're being built. Conformance with the state building code means that a house usually can be put anywhere in the state, since state codes override local codes, although there are exceptions such as in cities like New York City, Chicago, and Milwaukee.

The development of state-wide building codes is relatively new but, of all things, it is already reducing corruption in the building business. Crooked building inspectors, long a plague, can no longer shake down a home manufacturer by claiming that his houses do not conform to the local building code. Often a phony charge, this has long been used to extort money from builders; no payment, no acceptance of the house.

Home manufacturers can now fight back. A New England manufacturer for example, told me of this problem with houses he shipped to a Boston suburb. A building inspector, seeking a payoff, wouldn't okay the houses even though they conformed with the Massachusetts state building code. The manufacturer appealed to the state code officials. The local inspector was firmly told to shape up. The houses met the code and he could not disallow them.

Sizes and Prices

Modular houses range from modest one-story models with 1,000 square feet of floor area, more or less, up to large models three to four times that size, which is a lot of house. That means finished houses valued from as little as $25,000 to as high as $150,000. Actually, there's no limit to the size and price of modulars. Theoretically, sections can be stacked vertically, as well as horizontally, which makes available a large variety of house sizes and shapes practically at the push of a button.

A growing number of multistory apartment houses and condominiums are, as a result, also being built from factory-made modular sections. Again, most people seeing them don't realize that they're looking at manufactured housing. One of the most famous examples of modular housing and the vistas it opened up is the much-publicized Habitat in Montreal, Canada, designed by Israeli architect Moshe Safdie, and unveiled at the 1968 Montreal Exposition.

Famed Habitat housing complex in Montreal, designed by Israeli architect Moshe Safdie, is the classic demonstration of modular housing. Individual sections were built in a nearby plant, trucked to the site, and lofted into place by a crane. Moshe Safdie & Associates.

Design and Floor Plans

This brings up a criticism of the modular house: not enough design variety. Part of the trouble arises because individual house sections are made no wider than 12 or 14 feet. Depending on the state, that's the maximum width or "wide load" permitted for house delivery trucks on the highways. At last count, all states except California allowed 14-foot wides; California * allows only 12 feet. Even if shipped by rail, trucks are required for final delivery, so no factory house sections wider than 12 or 14 feet are usually made. One exception is certain western states where full-width houses are commonly shipped short distances from factory to site.

Room sizes are often no wider than these dimensions, though the houses may be two to three times the width of each section. That's done by assembling two or three rectangular sections side by side. The total size and number of rooms is determined by the length of each section, which can range up to about 60 feet. Two 14-foot sections, side by side, each 70 feet long, can produce a good-sized house of 1,960 square feet. A house twice the size, or nearly 4,000 square feet, can be made by stacking two more sections on top for a two-story house. Because some highway rules mandate that factory house sections be no more than 12 or 14 feet wide, does not, however, downgrade the architectural design and livability of a house. As noted later, even a maximum room width of only 12 feet adequately meets the minimum architectural standards for good design of every room of a house.

There's also good news on the floor-plan front. Modular house designers are beginning to feel their oats and break through old barriers. Floor plans are being varied by adding a third or fourth section at a right angle to the main house to produce an ell- or T-shaped plan. More and more houses wider than 28 feet are made with three sections, designed to give triple-wide houses. In short, the modular house may have had plan limitations, but new ones offer greater variety.

* Just before this book was published, California capitulated and now allows 14-foot wides.

Maximum Savings with a Modular House

Savings depend on the brand and those other variables, cited earlier, that affect prices. The largest savings possible with modular houses are less than the savings possible with mobile homes largely because modular houses are built to conform with tougher, more demanding building codes and modulars often require more expensive foundations.

The summary fact is that the modular house is probably the best, strongest and most efficiently made house in the world today. Anyone who doubts this should be reminded of those foreigners who are ordering U.S. modulars despite hefty cost increases for overseas shipment. The cost of all modular houses also could come down in the future. Savings are limited now because the number of modular houses being produced each year is relatively low. As new methods of making and selling them are developed and manufacturers swing into high gear, greater savings should inevitably follow. Saving the most money on a modular house still requires the same effort required for top savings when buying anything else, from steak to cars: shop. And, as mentioned above, some makers offer special savings by selling modular houses with incomplete interiors which you finish with your own do-it-yourself labor. The house price is reduced accordingly.

A Few Final Questions Often Asked

How much can a modular house be changed or modified to fit a buyer's wants? Not extensively and not necessarily for all modular houses, but often, yes. Ask the salesman. There may or may not be a charge.

Exactly how do you finance a modular house? Much the same way that any regular house is financed. Often the manufacturer or his local representative can recommend a mortgage lender who knows the house and will give you a mortgage. But you should still shop among other mortgage lenders, too. Compare the different terms offered. By the way, many home buyers overlook a major way to save on this score. They don't know that shopping for a good mortgage can be important, since mortgages can vary greatly in cost and content.

What about resale value? Like any other good house, modular house values should increase in the future as they have in the past, assuming that overall housing values continue to rise. But just as the best cars attract the highest resale value, the resale value of modular houses should rise higher than that of typical stickbuilt houses in the future. That's because of the higher quality of modulars.

5.
THE PANELIZED HOUSE

With the exception of mobile homes, more panelized houses are turned out by manufacturers every month than any other kind of factory house. More home buyers, therefore, are likely to buy a panelized house than any other kind of factory house. It's a house made by scores of different manufacturers. It comes in a wide variety of types, sizes, plans, and styles—from low-priced beer models, up to high-priced champagne models, and plenty of variety in between.

Panelized means that the complete walls of a house are factory-made in large sections, or panels, usually 8 feet high and up to 40 feet long. Sometimes the doors and windows are factory-installed in the panels, which is called prehung. The wall panels are designed to go up quickly, immediately after delivery, one after the other. Then the structure can be quickly topped with the roof and closed and locked up within a few days—sometimes by nightfall of the first day. Things still have to be finished inside, to be sure, but getting a house locked up that fast is one of the major accomplishments of factory-made houses.

Some manufacturers will also provide a panelized floor system or a roof system made in panels for fast installation, or both, supplied with their packages.

A panelized house may not save you as much time and money as a modular house. But for that trade-off, it offers these advantages:

1. Great variety of floor plans.

2. Most can be modified to suit a buyer's needs.

3. Savings to buyers, including professional builders and developers, who want houses made in a factory from their own plans and specifications. Many manufacturers will turn out one or many panelized houses for any buyer.

4. Savings in time and money for the do-it-yourself buyer who wants to save by doing his own building, but from a sophisticated house kit instead of going through the agony of building a whole house from scratch.

5. Savings on a vacation house that fits any of the first three categories just cited.

How Good a House Is It?

Like most manufactured houses, panelized houses conform with state and regional building codes. As a result, they are usually better-made than required by most local building codes. Like modular houses, the panelized house incorporates excellent workmanship and construction. There may be a few exceptions, perhaps, since not every apple in a barrel is perfect. Number One lumber and high-grade materials are used. Workmanship is usually excellent since it's a natural consequence of factory fabrication. Parts are put together in forms and jigs which lock things into place like the mold for baking a cake. Also, there's little room for sloppiness because of the quality control over the assembly-line production of panelized houses.

Good workmanship, by the way, does not always extend to the completion of houses at the site because of site construction being more susceptible to human errors. To avoid site mistakes, some manufacturers send their own factory crews to assemble their houses.

In general, this also means that the greater the number of house parts made in a factory, the less labor required at the site. Thus, the less likelihood of sloppy work and mistakes creeping in when the house is being finished. The on-site construction location of a new house is, like a battleground, a vulnerable place. The less action there, the better.

The Panelized Package

The package is known in the trade as "the wood." Its contents from the factory vary from manufacturer to manufacturer. In addition to panelized walls, it usually contains the prefabricated parts for all or most of the basic structure of the house, i.e., floor, ceiling, and roof. The outside doors and windows are supplied separately, if not installed in the wall panels. Some manufacturers provide little more to complete the house. The rest must be provided by the house buyer. Other makers provide more, including kitchens and bathrooms, heating, wiring, and plumbing.

One comparatively small manufacturer, for example, supplies panelized wall sections for its houses, but only precut parts for the floor, ceiling, and roof. The 4″ x 10″ floor girder, floor joists, and subfloor plywood panels are provided cut to size, but must be nailed together at the site. Windows and doors are supplied with the wall panel system but the interior partitions are made from bundles of precut framing supplied with the package. Other parts supplied with the package are: roof trusses, sheathing and shingles; roof flashing and gutters; stairs, including railings; hardware, which means nails, screws, plates, and bolts; kitchen and bathroom cabinets; and interior trim and doors. Cedar interior paneling and fireplace are optional. All else must be provided by the buyer.

The package made and sold by a large midwestern manufacturer illustrates a more complete factory house package. It is designed to cut building costs by sharply speeding up on-site erection. It provides the complete floor, wall, ceiling, and roof panel systems. The wiring and insulation are factory-installed inside the wall panels. Special kitchen-bathroom cores are made as a unit and shipped with each package. Much of the rest of the house is also supplied. A list of what you get in a well-furnished house package together with specifications is given on pages 56–59, in chapter 6. About the only significant difference between that package and what is supplied with a panelized house is that for the panelized house the walls, and

Here is how the shell of a panelized house is erected and closed up within a day or two. The exact shell completion time depends upon the size and type of house. The foundation, with or without a basement, is, of course, prepared in advance. After the house parts arrive, wall panels are erected around the perimeter of the house, roof trusses are installed and covered with panels, and the house is locked up. Kingsberry Homes, Boise-Cascade Corp.

sometimes the floor and roof systems, are made and supplied in large panels.

"Open" Versus "Closed" Panels

These terms have to do with the degree of completion of the wall panels for a panelized house. They are either open or closed. A house delivered with closed panels is more complete when it leaves the factory. Its panels are shipped containing insulation, wiring and electrical outlets, all installed in the factory, and are closed on both sides. The exterior wall siding (outer skin) closes the outside, and the finished interior wall surface (a skin of gypsum board or paneling) closes the inside. Once closed-panel walls are erected, usually in a day or so, the outer shell of the house is not only closed up but complete.

A house with open panels obviously requires more time and on-site work to finish. Open-panel walls are made only with their outside surface cover installed in the factory; the inside is open and hollow. Insulation, wiring, interior skin, and anything else needed must be put in the walls after the panels are in place. Only then can the inside of the walls be closed; that's usually with gypsum board.

A house with closed panels can save time and money, with one exception. Closed walls are made in the factory considerably faster and better. Fewer things are likely to go wrong in the factory, compared with the site where gremlins abound. The house with closed walls is, in short, more efficiently and economically made. Besides, the manufacturers who turn out closed-panel houses clearly strive for a more advanced and sophisticated product. This should be reflected overall in their houses being more advanced and sophisticated.

The exception is an open-panel house for the do-it-yourself buyer who wants to save money by using his own labor to complete his house. The extra time required to complete the house is less important. Squeezing globs of insulation inside the walls, threading the wiring through, and closing up the panels is the meat he feeds on.

The more, the merrier, since all the additional labor he provides is money in his bank.

The Wet Core

This is a mechanical core unit, or utility section, that contains the central plumbing, heating, and wiring equipment for a house all combined in one compact unit. Those are three of the most expensive organs of a house, especially when installed by three temperamental trades of workmen at the site. Efficiently made in a factory in one compact unit, the wet core is obviously a major step forward. It saves time and money. That's particularly true since this triple-threat unit can be quickly set in place, inside a panelized house at the site, with minimal opportunity for human error. It can mean major savings because it has been preengineered and made in the factory, compared with paying three different high-cost crews to install the same three components separately at a new house site.

Some manufacturers routinely supply wet cores with panelized houses. One main obstacle to its use in homebuilding has long been problems with varying building codes. This made it difficult for one standardized core to be supplied with houses shipped to areas with different mechanical code requirements. That's less of a problem now since state building codes have introduced the same requirements for all houses in a state. There has also been opposition to wet cores by plumbing and electrical unions that see it as a threat to union members. This is being at least partly overcome by having wet cores installed by union labor in union factories.

Buying a panelized house with a mechanical wet core is a good thing. I recommend it.

The Panelized House Summed Up

Compared with modular and mobile homes, a panelized house may save you less money but it can offer greater design variety. Because the panelized house is less completely made in the factory, it requires more on-site construction. But finding skilled people to complete its construction may pose a problem. This is easily overcome in an area where the manufacturer can supply a skilled factory crew to complete on-site construction of his houses. It can be more difficult in other places. Remember that completion of a panelized house in particular, requires erection crews with precision know-how and experience.

Compared with a precut house, on the other hand, a panelized house ordinarily can save you more money because it requires less on-site construction labor. This, though, can make a panelized house less attractive to the do-it-yourself buyer who welcomes a house kit that requires plenty of on-site labor. He fairly drools for every chance to use his own labor to save. A panelized house can still be his cup of tea. But all except the most skilled home buyers should probably step aside and let professionals erect the structural shell of a panelized house. And then, perhaps, step in and complete the rest of the house by themselves. Most panelized houses are, in fact, built and sold by manufacturers' builder-dealers, some on order from buyers, others built speculatively.

6.
THE PRECUT HOUSE

The precut house comes in a broad variety of sizes and types, and the quality of its materials can be quite high. But it is the factory house least completed in the factory. It is, in effect, the least educated factory house, a grade-school graduate at most. Because it requires the most on-site construction labor to finish after delivery, the quality of the finished house can't be guaranteed. It depends on how well its builder does his job at the site.

Also, because of the relatively large amount of on-site construction work it requires, it offers the least potential for buyer savings, with the major exception of the do-it-yourself buyer for whom the precut house kit glows with money-saving appeal. It is, in fact, the largest conceivable put-it-together-yourself kit of any kind available.

Offering maximum savings to the do-it-yourself home buyers is one of its main advantages. About the only other time a precut house makes sense is to obtain a special kind of factory house (log or dome home, for example) that is available only in a precut package. Usually the only alternative to get a special house like one of those is to build it from scratch.

The main drawbacks and limitations of precut houses? Two have already been noted: the comparatively small potential for saving both time and money it offered (again except for the do-it-yourself buyer), and it is more susceptible to on-the-job errors because so much of it is built at the site. A number of precut manufacturers do, however, provide traveling factory work crews to assemble their houses. Naturally, you pay their travel time.

The Package Ingredients

A precut house package, "the wood," generally contains all or most of the lumber for the main structure of a house, from floor to roof. The outside doors and windows are generally included. But after that there is no set type or quantity of package ingredients. It varies from barebones kits of precut wood on the one extreme, to bigger, more elaborate kits containing practically everything needed to build a house including the kitchen sink.

Here are examples: first is the barebones package sold by a New England maker, Timberpeg. It contains the precut lumber for the structural shell of the company's natural wood houses, which means floor, wall and roof system including shingles, insulation, windows, and exterior doors. Nearly all the rest of the house must be provided by the buyer, including foundation, basement stairs, interior partitions, all kitchen and bathroom parts and, among other things, the plumbing, heating, wiring, and lighting. Why buy such a package when the buyer supplies so much of the house? Because houses like this one can offer something special in their distinctive design, and the prefabricated structural parts supplied by the manufacturer do represent a major lift over building a new house from scratch.

The second example is a package with virtually all the ingredients needed for a house sold by such manufacturers as Capp Homes, Miles, and Ridge Homes. The Ridge package, for example, includes:

Floor system with steel bridging;

Wall system with prehung doors and windows (which means the windows and doors are preset in their frames);

Exterior siding (outside skin);

Gypsum board for the inside wall skin;

Roof-ceiling system including roof shingles, gutters, downspouts, flashing for the chimney, and plumbing vents;

A precut house comes with bundles of essential parts and lumber sized for quick nailing in place. New England Components/Techbuilt.

Interior partitions, doors, and closet shelving;

Kitchen cabinets, countertop, fixtures, appliances, exhaust hood and fan, and vinyl flooring;

Bath(s) medicine cabinet, sink, tub, toilet, shower stall, and ceramic tile;

Household hardware (door knobs, locks, and hinges);

Electric heating system, with a thermostat for each room;

Hot water heater, 200-ampere capacity, central electric board, and wiring package for all circuits;

"Rough" plumbing consisting of water lines for the kitchen and bathroom fixtures (but not the main plumbing supply and drain pipe system, which is one of the few major things besides the foundation not provided by the manufacturer). A full listing of everything supplied with this package is shown on pages 56–59.

In between those two extremes in precut house packages other manufacturers supply varied assortments of parts for their houses.

Obviously, the more complete the package, the easier and faster it is to complete the house. Time and money also can be saved when a house package includes materials like roofing shingles, insulation, and things like nails. A package with such sundries gives you the right amount of measured quantities of each for the house. Buying the same things from a local lumber yard could easily mean overkill, hence overpaying for things left over, also called waste. If in doubt about this, get bids from local lumber yards for the same materials, and compare them with the prices charged by the house manufacturer. This may sound like a big pain, but it can be quickly replaced by happy dollar savings as a result of shopping.

WHAT YOU GET WITH A FACTORY HOUSE PACKAGE

Exterior

Steel Adjustable Columns per Plan

Framing

Floor Joists—2"x8" or 2"x10" on 16" centers with steel bridging, ready to install for extra rigidity. 2"x6" foundation sill plates and 6"x10" built-up main beam.

Wall Framing—2"x4" on 16" centers. All corners include 2"x4" built-up corner studs. All window openings supported by two 2"x10" headers.

Wall Sheathing—½" thick 4'x8' plywood on all exterior corners, ½" asphaltic impregnated sheathing for increased structural rigidity.

Ceiling Joists—2"x6" on 16" centers pre-cut and beveled. 2"x4" stay board supplied for leveling, spacing, and stiffening each span of ceiling joists.

Roof Rafters—2"x6" on 16" centers pre-cut and notched, supported by a 2"x8" ridge board and tied together by 1"x6" collar beams.

Gable Wall—2"x4" on 16" centers with ½" asphaltic impregnated sheathing, 2"x6" plate.

Roofing

Sheathing—½" thick 4'x8' plywood.

Shingles—Asphalt shingles over 15 lb. asphalt saturated felt. Self-seal for extra weather protection with 15-year guarantee. Aluminum roof edging.

Gutter and Downspout—5" K-type aluminum (white) gutter with 2"x3" downspout and accessories.

Flashing—Aluminum flashing included for all valleys.

Exterior Wall Finish

Siding—White horizontal double 4" with aluminum continuous corners or vertical primed manufactured siding.

Trim—Aluminum (white) soffit, fascia and barge system. Aluminum soffit vents, rafter baffles and vinyl gable louvers.

Millwork

Doors—1¾" thick steel insulated, primed and prehung front and rear doors. Hardware and weather stripping supplied, drip cap where applicable. Aluminum threshold installed.

Windows—Primed white pine double-hung (or slider where specified) factory assembled with balances and weather stripping, ready to be set into the wall. Window and door blinds are included where specified. Caulking.

Storm Windows—Aluminum (white) combination storm/screen windows for double hung windows. Storm panels and screens for sliding windows.

Interior

Walls

Framing—All partitions, bearing and nonbearing are precut 2"x4" on 16" centers for 8 feet ceiling height with double top plates. Semi-assembled door bucks with two 2"x10" headers.

Insulation—R-11 rating for floors over unheated areas and exterior walls and R-19 for ceiling area. Sill sealer, box joist insulation, insulation at exterior corners and intersecting interior and exterior partitions.

Wall Covering—Polyethylene vapor barrier on exterior walls. Choice of ½" drywall board in 4'x8' or 4'x12' sheets complete with spackling cement, joint tape, and metal corner beads. In the bathrooms, your choice of colors in ceramic wall tile in the tub or shower alcove, 6 feet high from the floor.

Doors

Passage doors are semi-assembled prehung mahogany flush with hinges installed, center bored for lockset. Each closet has mahogany flush wood sliding doors.

Millwork

Trim—Solid, clear-grade, white pine casings for windows and doors, door stops, baseboard and floor molding. Pin rails, shelving and rod supports for all closets included.

Stairways—Basement stairway includes handrail, post, precut stringers, risers and treads. Where specified, finished stairway is preassembled, prefit, and predrilled (oak optional). It includes fir or yellow pine treads and risers, newel, handrail, balusters and other accessories applicable.

Flooring

Subfloor—½" thick 4'x8' plywood sheathing for strong floor platform.

Finished Floor—Choice of two styles of carpeting over ⅜" urethane padding over ⅝" particle board underlayment or $^{25}/_{32}$"x 2¼" #1 oak hardwood flooring, tongue and grooved and end matched over red rosin building paper. In kitchen and bath areas, choice of vinyl asbestos floor tile over ⅝" plywood underlayment or vinyl cushioned flooring over ⅝" particleboard underlayment.

Hardware

Nails—An adequate supply of nails of all sizes and types required in accordance with standard building procedures.

Finished Hardware—Brass finish door locks, hinges, door stops, sash locks, handrail brackets and aluminum closet rods.

Plumbing

Rough—"PVC" or "ABS"

plastic drainage system complete with fittings and ½″ M copper tubing water lines with all fittings.

Finish—Fixtures include a 24″ vanity with one-piece molded top, scalloped bowl and molded backsplash, plus water closet—in your choice of colors with a matching seat and cover. Family bathroom includes tub with shower; master bathroom features stall shower with tempered safety glass enclosure, plus water closet and 24″ vanity. Powder room includes water closet, wall hung lavatory and rectangular mirror medicine cabinet with bar light. Shutoff valves below all sinks and water closets.

Bathroom Accessories—Deluxe oval mirror/medicine cabinet and deluxe light fixture. Also chrome towel bar, toothbrush and tumbler holder, soap and grab bar, paper holder and shower curtain rod.

Heating

System—Electric baseboard heating system with individual room thermostats. System designed for local heating requirements.

Hot Water—52 gallon glass-lined, electric, quick-recovery domestic hot water heater with safety relief valve.

Electric

Rough—200 Amp service with switch-type circuit breaker panel. All necessary ceiling and wall boxes per the current National Electrical Code.

Wire—Wire package includes sufficient amount of 12/2 and 14/2 copper wire and breakers to complete all house circuits per plan. Also, proper size and amount of wire and breakers are supplied to install all appliances and utilities.

Finish—Your choice of several styles of lighting fixtures. All necessary receptacles, switches, plates and accessories are included. Bedrooms, dens and family rooms have switched receptacles. Door chimes provided. Smoke detector provided for each sleeping area.

Kitchen

Cabinets—Your choice of cabinets in hand-rubbed oak finish. All wall cabinets feature adjustable shelves.

Sink and Counter Top—Choice of stainless steel or colored single bowl sink with all fittings and faucet. The counter top is post-formed Formica in your choice of pattern and color.

Appliances—30-inch free-standing electric or gas range with removable oven door; dishwasher; M-4 ductless range hood with fan, light, and charcoal filter in a preassembled, prewired unit; hood.

Some parts supplied in a house package (kit) like this may be omitted or substitutions made, and sometimes additional parts may be supplied at the customer's option. Understandably, the specific products supplied will change from time to time. Though this shows what's supplied in a bulging precut house package, much the same contents may be supplied with a panelized house except that the walls and sometimes the floor and ceiling systems are made in the form of long panels, usually with prehung windows and doors. As noted in the text, there is an example of a fairly complete panelized house package. Many others come with fewer parts for the house and some with many fewer. Ridge Homes.

The Utilities

Plumbing, heating, and wiring kits supplied with some precut packages can simplify these three mechanical installations, but move cautiously here. For one thing, be sure that the plumbing and wiring will pass your local building code. Most do meet or exceed accepted national standards. The wiring with manufactured houses, for example, usually will pass the National Electrical Code. But that's not necessarily good enough for some building inspectors. A similar code conflict could cause a problem with the plumbing in a house package. Avoid this by checking on it beforehand. By comparison, the house structure itself usually meets or exceeds state and regional building codes and poses less of a problem.

For another thing, the decision whether or not to accept the optional plumbing, heating, and electrical wiring with a precut package can be influenced by the people who will install each in the house. If you must depend on local contractors for this, see them beforehand. Will they install the equipment supplied by the manufacturer? Contractors often buck at this because not supplying the equipment costs them the opportunity to make more money; in other words, the profit made on the materials as well as on the installation. And it's partly because some contractors are put off installing equipment that they're not familiar with, sometimes with good reason: they don't know how. If you're going to install such things yourself, taking what the manufacturer offers with his package can simplify things for you. If not, trouble and contractor anger can be avoided by being sure about the installation before ordering the mechanicals with the house package.

Optional Changes

The options offered with precut packages break down into two categories, much like those offered with panelized houses: design (change-of-plan), and material options. The first has to do with having a manufacturer add a room or two or an extra bathroom, say, to one of his standard models you like, making the house a couple of feet longer, or merely putting in a sliding glass door. Changes like that usually can be made though they may cost extra money.

The second, material options, have to do with your choice among the brand and color of the materials supplied with a house. Examples include a choice of outside wall materials, such as horizontal or vertical wall siding; different kinds of windows, kitchen cabinets, and appliances; gas, oil, or electric heat, and so on. There is usually no extra charge for these except for an option chosen that costs decidedly more than another; if it costs decidedly less, you should get a credit.

Another category of options is less well known. It is obtaining special-quality products, particularly those that will pay high dividends in savings and economy. These are things like specifying extra insulation and double- or triple-pane insulating glass (good to have in a cold climate), top-quality brand products like Andersen windows, a Kitchenaid dishwasher, two of the best for their respective uses, and an A. O. Smith high-efficiency water heater. Other high-quality products like these are discussed later.

Shipping Distances

Unlike other factory houses, some manufacturers will ship their precut houses long distances. Some makers ship their packages anywhere in the country and overseas, too. This can be done because the precut package contains less of a house than other factory packages, therefore weighs less and its long-distance shipping costs are not as high. This can let you buy a desirable house that is otherwise not sold where you live. As noted earlier, though, be sure that a house shipped beyond the manufacturer's normal marketing area will meet your local building code.

To sum up, the precut house is half a loaf or less because much of the material for it often has to be provided by the buyer and more of it has to be built at the site, compared with other factory houses. Nonetheless, the really difficult construction—the basic structure—is provided, and that's a major step forward compared with building a stickbuilt house.

The precut house can be a boon for the do-it-yourself buyer who wants to save a lot of money by using his own construction labor. Precut houses include handsome and well-designed houses and special houses that are unavailable elsewhere in the factory house field.

7.
LOG, DOME AND A-FRAME HOUSES

This is a trio of special houses that can now be had in precut factory-made packages. In other words, these houses are subcategories of the precut factory house. Anyone who still builds one from scratch, not from a factory package, should have his head examined.

Log cabins, chalets, and lodges have always been popular in Canada and have had a surge of new popularity in the United States beginning in the 1970s. Though widely used for vacation houses out in the woods, their natural habitat, more and more are now being built—by do-it-yourselfers as well as contractors—for year-round houses. And you don't have to be a pioneer to live in one.

They're made in a great variety of styles and factory-made with varied construction techniques. Though well suited for rugged backwoods terrain, they no longer demand rugged living within. They not only have electricity and central heat, if desired; some models are equipped with cathedral ceilings, porches, dormers, lofts, skylights, and even a matching doghouse.

Log houses seem particularly appealing these days because of our collective nostalgia for the way things were, because of their rustic looks and handcrafted appeal, because they're quick and inexpensive to build, and because they come in factory-made kits. Since they require neither framing nor siding, they go up quickly and require little in the way of maintenance—an occasional coating of preservative, indoors and out, as a substitute for paint.

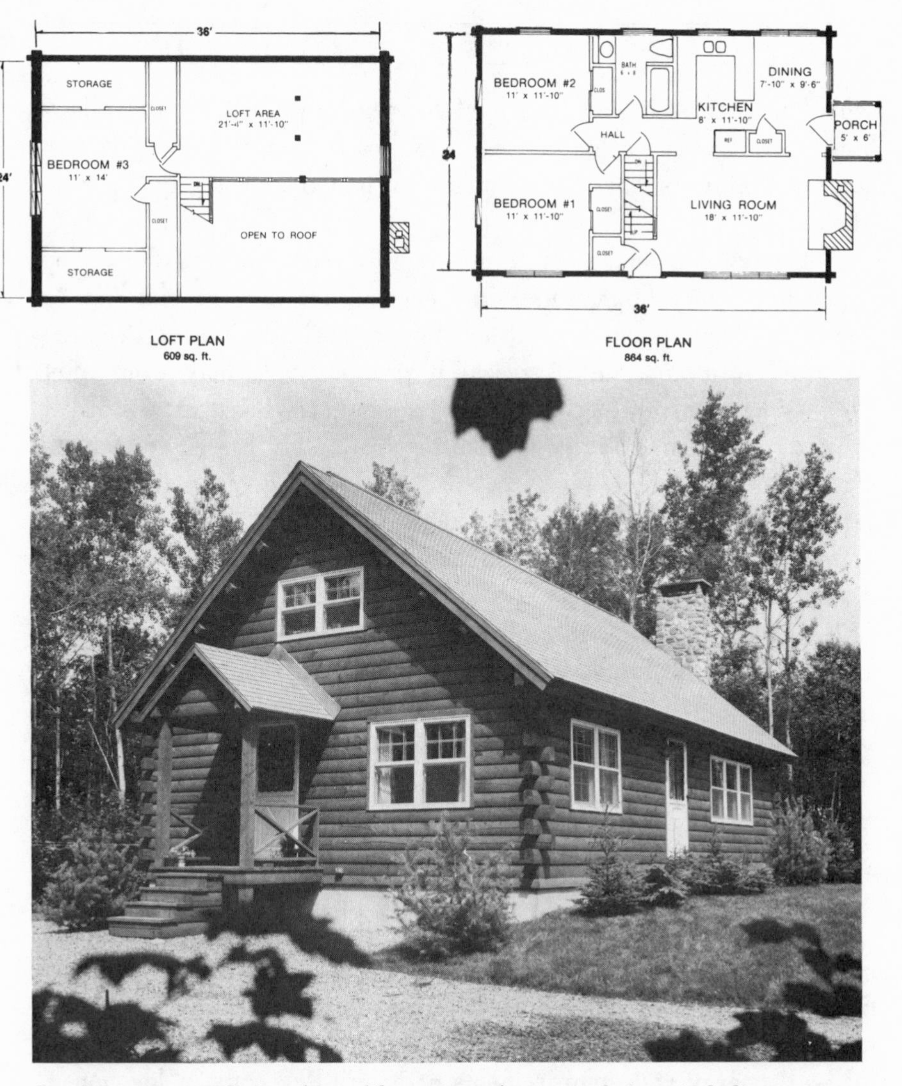

Log houses are made in a forest-like variety of types and sizes, including two-story models like this one. Northeastern Log Homes.

Unless you choose one of the few double-walled structures with insulation between the log layers, what you see on the outside is what you get on the inside: solid timber walls as thick as desired, or usually from 6 to 8 inches. Solid timber construction, moreover, is long lasting. The owner of one log-building school says, "A log house, if built correctly and preserved properly (given periodic coats of preservative), should last for 250 years." A log-building school is, just as it says, a school for teaching builders and others how to put together a log house. Many are inexpensive and give fast courses. Names of ones near you can be had from a log house manufacturer.

Though log houses offer most of the conveniences demanded by contemporary homeowners, the energy efficiency of many is "a sticky question," says Terry Sherwood of the U.S. Forest Products Laboratory, a government research group. Some may even offer fuel savings; current occupants of log structures boast of low utility bills. Supposedly, a log house can be adequately heated by a good airtight wood-burning stove, and log house makers claim that it will retain heat better and use less energy than an equivalently sized frame house. Most log houses do meet most regional and local codes. A number of them are also approved for Federal Housing Authority (FHA) and Veterans Administration (VA) financing. But in some areas uninsulated log walls may not meet energy-code requirements for insulated walls.

Experts point out that solid wood, as in log walls, is not a top insulator. Nonetheless, a solid log wall six-to-eight-inches thick is equivalent to an ordinary house wood-stud wall with up to two inches of insulation. That's not bad, considering that the first inch or two of insulation saves the most heat.

In addition, log house people say that the government energy criteria for houses do not give credit for such things as heat absorption (from the sun, for example) where log walls do well. And even though uninsulated log walls of a log house, per se, may not meet the wall energy code requirements of a building code, a log house, nevertheless, can still meet the overall energy standards for whole houses. The log people accomplish this by beefing-up other portions of their houses—roof, window design, etc.—with more insulation and other energy-saving construction. That keeps down the total energy loss of the house. As a result, a well-designed log house can meet the national energy standards for whole houses, such as that of the highly regarded ASHRAE standard (American Society of Heating, Refrigeration and Air-Conditioning Engineers). The acid test is that then log house fuel bills will be low.

In some houses, the makers solve the problem by slipping insulation in the log walls. A variety of other solutions are offered by different manufacturers to produce a high enough R value to meet stiff code requirements. One maker's logs are hollowed out; the dead air within allegedly insulates better than solid timber, though this could be questionable, too. Other logs contain insulating fiberglass, urethane, or other such materials. Some logs are joined with insulation or gaskets that form airtight, watertight seals. Some log walls are built with nails or spikes, some with splines. Some use traditional chinking. Some require caulking. Some combine any number of these methods. And some manufacturers' precision-cut logs fit together so tightly with saddle notches or tongue-and-groove construction that the blade of a knife cannot pass between them.

Whatever system you may choose, there are other choices to make as well. The timber used can be cedar, pine, fir, poplar, spruce, or other types. Northern white cedar is the gem of the bunch and is the best insulator. The U.S. Forest Products Laboratory rates cedar, spruce, pine and fir, in that order, as the most desirable woods for log houses. Logs are either hand peeled or machine peeled, air dried or kiln dried. They may be left rounded, flattened top and bottom, or hollowed to fit onto the adjacent logs. Most, however, have grooves to accommodate electrical wiring. Some are dipped in preservative to repel rot, mold, insects, and rodents; cedar by itself has natural resistance to bugs (including termites) and to rot.

With any construction system, precut log homes in kit form can be erected by the buyer or by a local builder. Some manufacturers ship only logs cut to conform to the buyer's blueprints. Most offer many models, all of which can be modified if desired. In general, even the most expensive log cabins will cost less than a standard wood-frame house. Because of the variety of packages and options available, cost comparisons between log-house kits are hard to give. While the price of a kit-built log shell is invitingly low, construction costs can double or triple the bill. For example, one of the most imaginative log houses is also among the most expensive. In 1979, a kit for a highly energy-efficient three-story round house made by Lodge Logs of Boise, Idaho, cost about $15,000 and included double log walls with urethane foam insulation for a very high R value of 35. Also available with it, for about $11,000, was a solar collector and

A dome house means life in the round with windows located almost anywhere you wish. Cathedralite Domes.

heating system run by two minicomputers (whatever they are). (As noted in chapter 13, buying mechanical solar heat today is buying a pig in a poke, and in addition you almost always need a regular heating system too, for backup. It may be wise to let others be the guinea pigs. These are, however, other solar heat methods that are indeed worthwhile, also noted in chapter 13.) Depending on location and interior trim, total building cost for this house could run, in 1980, from $60,000 to $90,000.

Factory packages don't include the foundation, and rarely the roof, flooring, heating, plumbing, or wiring, so finishing the house adds considerably to the cost. Some kits include doors and windows, while other manufacturers advise buying these locally. Labor is obviously an important factor in determining the construction cost of a house. Doing the major construction oneself can save between a quarter and a half of the finished value of a house, says one industry expert.

To lower your building cost, look for a good company with a nearby factory. After all, the shorter the distance the logs must travel, the lower the shipping expenses. Determine how much of the building you can do and how much help you'll have. Then figure the contracting cost of having the foundation built, the walls installed, and so on. Find out which components are cheaper locally and which should be ordered from the manufacturer. But before you start, of course, check that your local building code permits a log house and arrange for the financing. A house that you can neither pay for nor live in is no bargain. A surprisingly lot of supposedly intelligent people have had to find this out the hard way.

Up-to-date facts about log houses are reported every year in the "Log Home Guide for Builders and Buyers," available for $7.50 from Muir Publishing Co., St. Frances House, Haliburton Lake Road, Fort Irwin, Ontario, Canada, KOM *1*PO.

The Dome House

This is Buckminster Fuller's geodesic dome, designed to stretch a minimum of material over the maximum area, and it is one of the strongest structures ever built. On a cost-per-foot basis, it could be the cheapest, too. Moreover, a dome is adaptable, easy to build, and inexpensive to live in. But though this geometric masterpiece provides unique shelter, it also demands a unique lifestyle, and it could keep the wrong owners running around in circles.

The traditional geodesic dome is composed of flat plane triangles bolted together to form curved pentagons. Because of the inherent strength of the triangle, it is ideal housing in an earthquake or tornado area. The wooden or metal framework is covered with either a canvas or plastic skin or with wooden panels. The structure is self-supporting; no interior walls or posts are required (though load-bearing walls must be added to support a second or third story for sleeping lofts).

The house may be erected on what's called a riser wall, or it may rest on a foundation, with or without a basement. As many as ten skylights are incorporated in some models, and five trapezoidal areas around the base can accommodate windows or doors; two or more domes often are attached along these openings.

Sizes vary from small one-floor dome houses of about 26 feet in diameter to large, elaborate houses with two or three levels and a diameter of 45 feet. The 35-foot diameter dome contains 1,740 square feet of living area over two stories. The 39-foot dome, perhaps the most common, offers 2,200 square feet, or a lot of house, over two levels and is 16 feet high at its center (the best location, by the way, for a fireplace).

Depending mainly on size, the dome shell by itself will cost up to about $15,000, which represents about 25 percent of the finished cost of a complete dome. Sold nationally through builder-dealers who may or may not participate in construction, domes are 5 to 15 percent cheaper to build than conventional houses. One builder claims it costs 8 to 10 percent less per square foot to build. In 1979, dome costs ran about $30 per square foot. A simple structure was only about $25 per square foot, while a more lavish dome could be about $34. A complete geodesic dome house can cost as little as $25,000. But a special exterior adds to the cost, and many buyers demand extras like decks, dormers, and additional skylights to lower lighting costs. Fuel bills, by the way, can run 30 to 50 percent less than those for heating a conventional house of the same size; because there is less surface area in a dome, hence less heat loss.

Of course, a do-it-yourself buyer can cut the cost considerably, as with other house packages. One dome manufacturer estimates that 75 percent of his customers do all the work on the shell, and 25 percent do almost all of the rest, too. Though an unskilled buyer is not advised to undertake the foundation or install the roofing, plumbing, wiring, or the heating, the shell construction is relatively simple even for amateurs. The mass-produced parts are designed for on-site assembly, and a whole dome can be completed in one to three days after the foundation is ready. Some kits contain preassembled panels to expedite work.

As mentioned earlier, the price of a dome kit, or package, depends on various factors, including distance from the maker's plant, its size, and the materials included. Besides the framework, most packages include the skin and some provide the wall-roof insulation. One manufacturer offers panels that incorporate the outside wall skin, insulation, and wood-paneled interior walls. Doors, windows, and the triangular sections called wings made to flank them, are usually included. Some makers offer canopies and other light openings such as transoms and skylights. Occasionally, interior paneling is included and roof sealant almost always is.

Sealant raises a major problem that can confront dome-owners—leakage. Though various companies provide various means of waterproofing, it can be less than efficient. Thus it rains indoors when it's raining outside, an unwelcome happening. Tar and webbing over seams can work. But for assured nonleakage, a regular roofing cover like asphalt or fiberglass shingles, wood shingles or shakes is recommended. One of the newer dome-roofing methods is a spray-on insulant covered with a rubberized roofing material. Finishing the interior can be expensive because most commonly used materials are geared for rooms with right angles.

Other drawbacks to dome living cited by some people can be considered pluses by others who like life in the round. Decorating offers interesting problems; if you want to hang a picture on a wall, you may have to add the wall. But since there is so much flexibility in the division of space, you have virtually total control over the indoor environment (but the fewer partitions added, the less air-space patterns will be blocked). Acoustics within a dome are marvelous, which is wonderful for music appreciation, but perhaps less wonderful for privacy.

As domes grow more popular, the nitty-gritty of codes and financing is easier to deal with, especially since some domes are approved by the FHA and the VA. Still, they do present the problems commonly encountered with any new and different kind of housing. In fact, a dome might meet even more resistance than a log house because it is newer and even less conventional. But building codes should not be troublesome, particularly if the structure is professionally built. Some manufacturers provide engineering drawings to demonstrate the structural soundness of their products. Lending institutions probably won't be difficult to convince, either, but obviously you must check before ordering a dome.

The A-Frame House

Initially designed for the cold north, particularly in ski country, A-frame houses were to the 1950s what the dome is to the 1970s: an innovative, inexpensively built form of architecture suitable for those who don't mind—or emphatically desire—an open, airy, contemporary house. Like the geodesic dome, the A-frame yields strangely shaped living quarters without the usual complement of right-angled walls.

Viewed head-on, an A-frame structure looks like what it's called, a letter A; or an inverted V. As in a dome, roof and walls are one and inseparable, and the interior space is open and airy. The steep roof, usually covered with roof shingles or shakes, makes the A-frame ideal in heavy snow areas; snow slides right off. The angle of the roof also minimizes wind damage, leakage, and maintenance work, and results in a longer house life.

An A-frame house is inexpensive to build because the strong roof members double as the wall framing. To cut costs further, this style is available in modular form. One manufacturer offers a modest single-wide A frame, 13′ x 36′, thus 468 square feet of living space. Two double-wide designs, 22′ x 36′ and 22′ x 40′ give you 792 square feet and 880 square feet respectively. The square-foot areas just given are for ground floors. Each can be had with a second story and an optional basement can provide even more room. An A-frame house designed for vacations and weekends, as so many are, frequently offers a cathedral ceiling and a sleeping loft that occupies only half the upper space. The A-frame house can not only be an exciting place that is surprisingly airy and open, but it also feels twice as large as it actually is.

Total cost of an A-frame house should run about 10 percent less than a house with conventional walls and roof. The largest construction savings, of course, are due to consolidating the walls and roof,

The Swiss chalet is the forerunner of this A-frame house. American Timber Homes.

and the simplified interior. As in nearly all other houses, however, plumbing is plumbing, heating is heating, and wiring is wiring, so the cost of these and other such components is just about as much as in a conventional wall-roof house.

Like a dome house, an A-frame should be insulated to the hilt. Its roof-wall structure usually needs as much insulation as the roof of a regular house. And its doors and windows should be equally well insulated.

8. MOBILE HOMES

Forget everything you've heard about mobile homes, start fresh and take them at their own terms. They have many faces and more and more are surprisingly attractive.

True, there are still many poor cousins among them that are far from good looking, and some of the places where they stand are little improved over the tawdry trailer homes of yesteryear. And many a new mobile home park, the new term for such neighborhoods, are as attractive as any new residential area.

More than that, mobile homes have now made a major breakthrough. New house developments in a growing number of residential areas are being built with mobile-home houses, though only experts can usually tell that they are mobile homes. The houses are designed and built like conventional houses. They're made with conventional wood-siding walls, asphalt shingle roofs, conventional windows and doors, and virtually everything inside conforms with standard home-building practices. And they are located on permanent foundations. The big difference, however, is that these "mobile" homes are priced sharply lower than conventional houses.

Two of these new breed housing developments were mentioned in chapter 1. One significant new one, a pace setting development, is Lake Mountain Estates, overlooking Lake Mead in Boulder City, Nevada, some 25 miles from Las Vegas. This is an area where nearby

luxury houses at the time were selling for $125,000 and up, according to *Housing* magazine, a trade journal.

The houses in Lake Mountain Estates, basically double and triple-wide mobiles (see photograph), sold for $39,000 up to $85,000 including land. Each house is built on a permanent foundation, with garage and landscaping, and the average cost of the land came to $17,000 per house.

One couple nearby seized the opportunity to make a short move for a big profit. They sold their $175,000 conventional house and bought one of the "mobiles" in Lake Mountain Estates for $68,000. The new house is hardly a come-down. It is roomy and satisfactory for, as you can see, considerably less money.

The ability of mobile-home manufacturers to make attractive houses at sharply reduced prices could have significant meaning for all home buyers. The only other houses close to meeting their prices are modulars. The mobile-home makers are also using their experience and mass-production techniques to turn out "sectional" or conventional houses that meet conventional building codes and are

A well-designed mobile home like this can hold its head high in any neighborhood. Reprinted from Better Homes & Gardens.

And so can this kitchen in a mobile home. Manufactured Housing Institute.

financed by conventional home-loan mortgages. Good looking, low-cost, one-at-a-time mobile-homes also are obtainable for individual home buyers who want houses on their own lots.

It should be emphatically mentioned that such topdrawer mobile homes may still represent only a minority of all mobile homes. Unfortunately, too many mobiles are still losers in the looks department. But 300 hitters are not numerous in the big leagues, either. It's easier, however, to upgrade mobile homes, so the number of good ones that become full-fledged members of residential America should increase. They could leap forward in popularity, considering the big savings that they offer.

The Advantages

Mobile homes shine in these ways:

- Their cost, less land, is almost half the price of conventional stickbuilt houses put up by local builders. As mentioned in Chapter 2, in 1979 the price for the average new stickbuilt house, less land, was $49,300, or $29 a square foot. The same size mobile home (1,710 square feet) sold for $26,350, or $15.50 a square foot.
- Built to order, mobile homes can be delivered and ready to move into (completely furnished, if desired) in as little as six weeks from the time of order, and sometimes faster. That compares with at least three to six months for a stickbuilt house to be ready.
- All mobile homes built after June 1976 must conform to U.S. government standards for construction, fire, and electrical safety. All must be equipped with emergency exits, such as pop-out windows, and smoke detectors.
- Interior and exterior maintenance is minimal. Many mobiles have tough, specially treated exterior walls that need only to be washed with detergent or antioxidant.
- Most mobiles are built in units 12 and 14 feet wide, and up to 75 feet long. Some are built as "left-" and "right-hand" units that are bolted together at the home site. Together, the two section units provide up to 2,200 square feet of living space or nearly 30 percent more than the average new U.S. house.
- Mobiles are offered with different floor plans, a number of which

can be modified to suit individual needs. Though many are locked into the elongated "Pullman car" pattern, manufacturers are breaking out of this monotony and offering refreshingly new room layouts. These include new H and U plans that are called tri-wides. They're made with two-section mobile homes separated by a central room in the middle (H plan) or at one end (U plan).

In addition, special add-on sections can add needed floor space as well as welcome variety to a mobile-home plan. There is a trio of them to choose from. The tip-out is mainly an alcove, usually 4 to 5 feet deep and up to 12-feet long. It folds, accordion style, into the mobile-home section for delivery, is pulled out to its full-size glory at the site. The slide-out (also called pull-out) is similar but larger, thus a more spacious alcove.

There is also the tag-out, in effect, a whole additional room. It's available in varied lengths, 32-feet being very common. The tag-out is more flexible in that it usually can be bought later if desired, while tip-outs and slide-outs usually must be ordered when you buy the mobile home.

• More and more mobile home parks are attractively set amid tree-lined streets and have social activity centers and swimming pools. Many are located near golf courses. Although their mobile-home units are placed comparatively close together—on lots of less than a quarter acre—privacy is enjoyed by virtue of careful landscaping, shrubbery, and privacy fencing. Roughly half of the close to six million mobile homes in use in the United States in 1979 were located in mobile home parks and subdivisions; the other half were on private lots. Some parks are for "adults only," with one person in a couple required to be at least 52 years old, thus assuring peace and quiet for those who are bothered by children.

• Because of mobile-home design improvements, many cities that once banned them now allow them on private lots.

• More and more mobile homes, especially those located on permanent foundations in attractive settings, do not depreciate markedly. Many are going up in value. This trend has been accelerated by the development of parks of condominium mobile homes. In one near Chicago, individual homes and lots sold for up to $31,000 when the development opened in early 1977. Some two years later, the same homes and lots were selling for nearly 50 percent more, or up to $45,000.

• Most mobile homes continue to be financed with personal loan contracts like an auto. But if a mobile home meets certain specifications for size and construction set by the FHA and the VA, it can qualify for the same FHA- or VA-guaranteed mortgage loan used for conventional houses. This can mean a low down payment, as well as a lower interest

This three-bedroom mobile home can meet local building codes. It was developed by Family Circle *magazine and the Armstrong Cork Company, for production by the Vindale Corporation, Dayton, Ohio, and Golden West Mobile Homes, Santa Ana, California.* Armstrong Cork Company.

rate. One requirement calls for the home to be on a permanent foundation.

Prices

In 1979 the average price of mobile homes was just about $15.50 per square foot of floor area. Thus, a single 12-foot wide unit, 70 feet long cost $14,280, more or less, depending on the brand and special features. A single 14-foot model, 70 feet long cost about $17,000, again depending on the particular model.

Double-wides cost up to twice as much, depending on the length or, in other words, the total enclosed floor space. For example, a generously large mobile home with three or four bedrooms and up to 2,000 square feet of living space can be had for $30,000 to $40,000. The going price for conventional stickbuilt houses of the same size was up to $70,000.

Another 15 percent of the price of a mobile home must be added to cover what's called the set-up charge. That's for tying down the home, (storm holders, etc.) and such things as skirting, porch steps, and minor landscaping. On the other hand, sometimes a thousand dollars or more can be saved by ordering a barebones, or unfurnished, mobile home. Not everyone needs the furnishings, including pictures on the walls, that usually come with the typical unit. You may have your own furniture for a house, or you may wish to buy your own on your own sweet time.

In addition, the furnishings often provided with most mobile homes are not known for quality, to put it kindly. To put it accurately, the typical furnishings provided are more like the low common denominator products and materials provided in typical stickbuilt development houses. Which is to say the lowest cost and quality that the builder, or in this case, the mobile-home maker, wishes to get away with.

There is also the cost of the land, as there is with any other house you may buy. If you put a mobile on your own land, the total land cost obviously must include the cost of excavation, foundation, paving, and utilities, as well as the raw-land lot cost. If you move into a mobile home park, you will generally pay a monthly ground rental fee, which varies from park to park.

Shopping and Buying Tips

A particularly good way to shop the market and see many of the different mobile homes available, all at one crack, is to visit a trade show. You can see virtually the whole spectrum. Shows are held regularly in nearly every part of the country. A local mobile-home dealer can tell you where and when the nearest ones to you are held.

The big national mobile-home show is held each year in January in Louisville, Kentucky.

Next best eye-opener is to visit mobile-home dealers. See at least a few. Before you buy, shopping around is essential not just to save money, but also to help you learn about special features to have, and on the black side, not to have. Dealers like to unload many extras on buyers that you can do without at good savings, and no thank you.

A dealer may say, for example, "Gosh, as long as you folks are here, take a look at a real beauty—came in just last week. It's all hooked up and, why, you could even move in tomorrow!" The unit the dealer shows is often loaded with what the trade calls flash, or more gimmicks and useless accessories than the jewelry counter at the five-and-dime store. The list price of every one is included in the inflated total price for the mobile home. Such questionable items include things like stained wood wine racks, flashy chandeliers, fancy lamps, large floral displays, as well as pictures and prints on the walls. The price of every such item is tacked on to the price of the house, unless you put your foot down and say no. Request a breakdown of everything you're being charged for and cross out what you don't want or need. That can save you good money.

Before buying a mobile home, check on the dealer by calls to the Better Business Bureau and a local bank or two that finances mobile homes. This smacks of the same stock advice for buying many other things, and it is. But obviously the greater the purchase money at stake and the greater the importance of a home to you, the more urgent the need for these calls when buying a mobile home—or any other kind of house. Do it and you will join the small club of people who benefit as a result, i.e., don't be one of the greater number of people who never make such calls and often pay the piper later in terms of problems and complications that could have been avoided.

If you are one of the half of all buyers who will live in a mobile home park, there's a good way to learn whether or not you're moving into a nice place or a Pandora's Box: *talk to mobile owners who live there.* Walk down a few blocks, ring doorbells and ask, like a survey taker. Say you're thinking of moving in and often it's like turning a faucet on. Facts about living in the place will come pouring out.

Be sure to ask specifically what's good and what's not good about the place and the landlord. Also, were any promised amenities or services, like landscaping and trash pickup, not provided? Are repairs made promptly? How many ground rent increases have there been in recent years? Is the park kept clean, safe, and quiet?

Most mobile parks are rated in "Woodall's Mobile Home and Park Directory," a yearly guide sold on newsstands. It's excellent for finding a good mobile home park, as well as providing late news and rating data about mobile home parks in general, as well as about specific ones.

If you are one of the other half of buyers who plan to put a mobile home on a private lot, check first with the local planning board and building inspector. This is important. Many places flatly outlaw mobile homes, though such bans are gradually disappearing as improved mobiles hit the market. Other places may permit them but local enforcement officials can make it tough before approval is given. Checking ahead of time is essential.

Checking Mobile-Home Quality Before Buying

An X-ray view into the anatomy of mobile homes can be had by visiting a factory and seeing how they're made. You'll see the different kinds of products and materials that make up the bones and body, the various organs, how the skin including insulation, is put on, how the kitchen and bath are made, and such things as how carpeting and other furnishings are installed. That anatomical course in mobile-home design and construction can be eye-opening. It can make you sharply aware of how to check mobile homes when you shop for one. Incidentally, mobile homes are seldom shipped more than 250 or so miles from the factory, so don't waste time considering models made farther away; they are probably unavailable to you.

Construction Quality

Because every mobile home made after June 15, 1976, Magna Charta day for mobile homes, must conform to the structural standards of the U.S. Department of Housing and Urban Development (HUD), a notice stating this should be on a mobile home you buy. Any dealer can show it to you. If not, he's not much of a dealer or he may be trying to pawn off an obsolete unit on you. Leave at once.

The standards not only specify the minimum requirements for the construction of floors, walls, roof, insulation, and other such structural shell items, but they also call for additional such needs. Examples are: two exterior doors remote from each other; at least one egress window in each sleeping room, i.e., a window that a person can get out of in an emergency; smoke detectors wired to the electrical system with audio alarms outside of each bedroom area; a tie-down system of cable straps that will make the unit hurricane-safe by anchoring it to the ground; and an electric wiring system that conforms to the National Electrical Code C-1 (which is the same code used for conventional houses).

Mobile homes are now, by the way, considerably less of a fire hazard than formerly when many inflammable plastics, including vinyl materials, were used in them. Because plastics are made from petroleum, the sharp increase in the cost of crude oil since the energy crunch began in 1973 has made these synthetic products costly. As a result, many makers have gone back to the use of wood paneling, plasterboard interiors and other such materials for mobile homes.

Buying a High-Quality Mobile Home

This can be summed up in one sentence: buy a top-of-the-line model. I mentioned earlier that marginal, or low-quality products are commonly used in run-of-the-mill mobile homes. These same mobiles, the lowest-cost ones, can still display the seal showing conformance with the government's national standards for two reasons.

For one, the HUD standards, like any other building code, do not cover everything that goes into a house. It covers only those parts of the structure that can affect the health and safety of people. For example, the HUD rules specify the minimum lumber size to make the floor, walls, and roof safe and strong enough at all times.

But they say nothing about the kind of interior wall paneling, kitchen countertop, or bathroom fixtures, among other things, that are necessary to retain their good looks, that are easy to keep clean, require little upkeep and service and will not quickly wear out. These features depend on the quality of such building materials and products, but have little or nothing to do with health and safety. Therefore they are not covered by HUD's rules or by other building codes.

In similar fashion, a heating furnace must be strong and safe and not be a fire hazard. But how well it is made for low fuel consumption, maintenance-free operation and long life—all having to do with good quality—are also matters not covered by the HUD standard or, again, by any building code that regulates house construction. Such extra quality is often lacking in most house products.

The second reason is that the HUD standard, a very good thing nonetheless, is still no more than a set of floor, or *minimum*, requirements for safety and health. If you want a mobile home that is better than minimum, obviously you must get one that is designed and made necessary to pass minimum structural standards. This principle, by the way, also applies to all other kinds of factory and stickbuilt houses.

In short, the typical mobile home that meets the government's mobile-home standards is a perfectly good product, but if you want a really good, if not top-notch, home, you must aim higher. That's to get a mobile home that will mean easier maintenance, lower upkeep, lower-than-usual monthly energy bills, good looks that will stay good-looking for a long time, plus a few other things that make the difference between dining on ordinary dinner wine and, instead, imbibing from a vintage bottle.

To get a first-rate mobile home, you must step up to a manufacturer's high-quality or top-of-the-line model. It's as simple as that. Of course, you will pay more. But the extra price is surprisingly little more. Many makers make at least two different kinds of mobile homes. The first is their standard or basic model that meets the HUD national standards. It's a good home but still a bottom-of-the-line mobile, designed to be sold at the lowest possible price. The second is their so-called "quality," or top-of-the-line model.

A top-drawer mobile home is better throughout because of better design, construction, and workmanship. Compared with the standard, lowest-priced mobile home, it has thicker and better wall paneling, mitered trim, 2″ x 4″ wall studs instead of the usual 2″ x 3″ (which may not be visible but which you can tell by a few raps on the wall), better carpeting with good underlayment; larger, better water heater; better furnace, and, on the outside, good long-life walls, rather than the cheap siding found on the lowest-cost mobiles. Other differences in quality between top- and bottom-of-the-line mobile homes can be determined by obtaining the specifications for each kind of home and comparing the differences.

Manufacturers of high-quality mobile homes include Dualwide Homes, New Yorker Homes, Ramada, and Skyline. Remember though that, like other products, including cars and appliances, some manufacturers make both low-priced, lowest-allowable-quality products, and much better-made models of the same product. So just buying a mobile home made, for example, by one of the companies just cited above, is not necessarily enough. Be sure that you get the maker's top-of-the-line mobile home.

The Warranty

Manufacturers sell mobiles with full or limited warranties of one and two years. Some cover only the structure and such things as wiring and plumbing. Others cover the appliances, including the furnace and hot water heater, dishwasher, clothes washer and dryer. If there's a product breakdown during the warranty period, service and repairs are provided by the manufacturer through the dealer you bought the unit from. Before buying a mobile home, you should obviously check on the warranty period and its small print.

Buying a Used Mobile Home

It's like buying a used car or an old house. Shop carefully and you can often find an excellent one at good savings. Let down your guard and you can end up with a lemon. If the previous owner maintained it well, the chances are then you won't get stuck. Inspect it carefully,

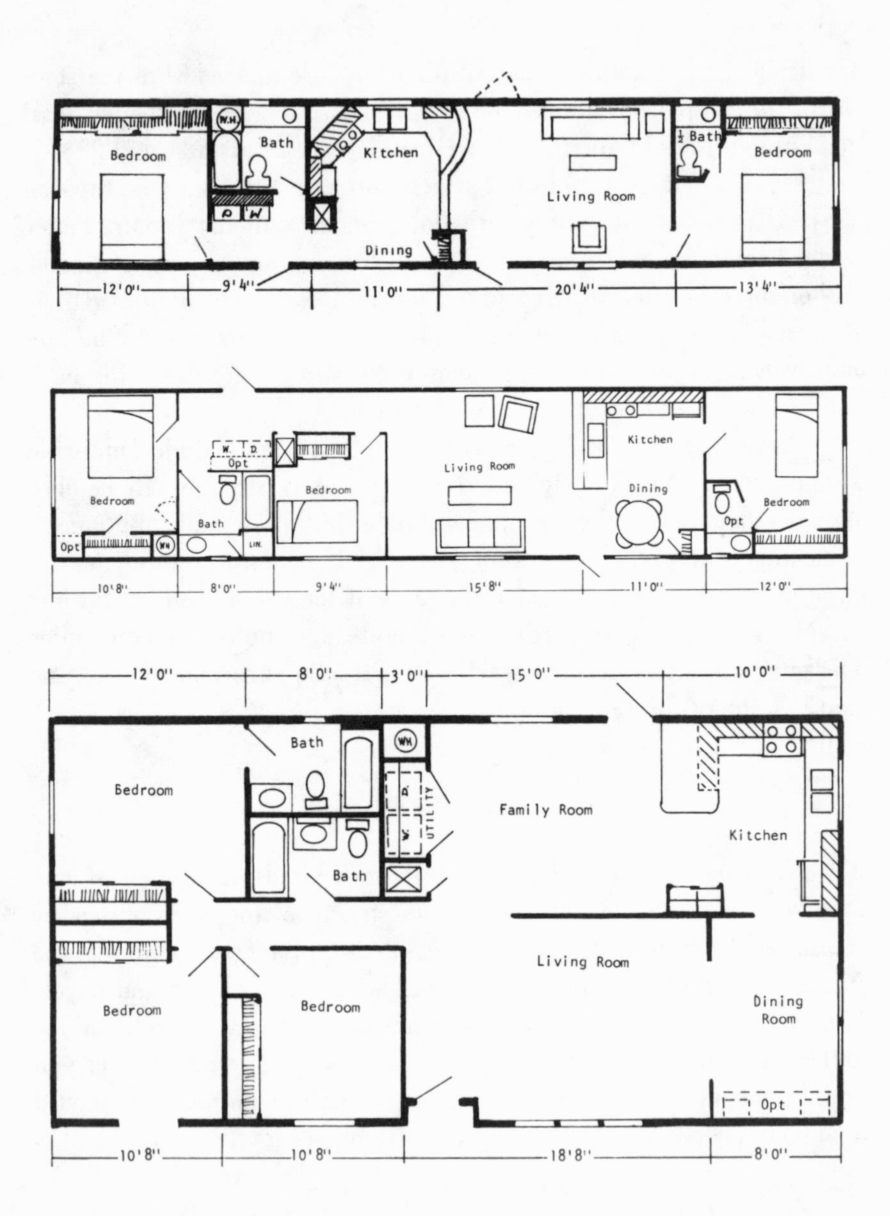

though, and be sure it is in good condition. Better still, hire an expert to inspect it for you. The small fee for this—perhaps $75—can be well worth it. To find an expert, ask people who fix and service them, or a banker who finances them.

Buying a used mobile can be better than buying a used house

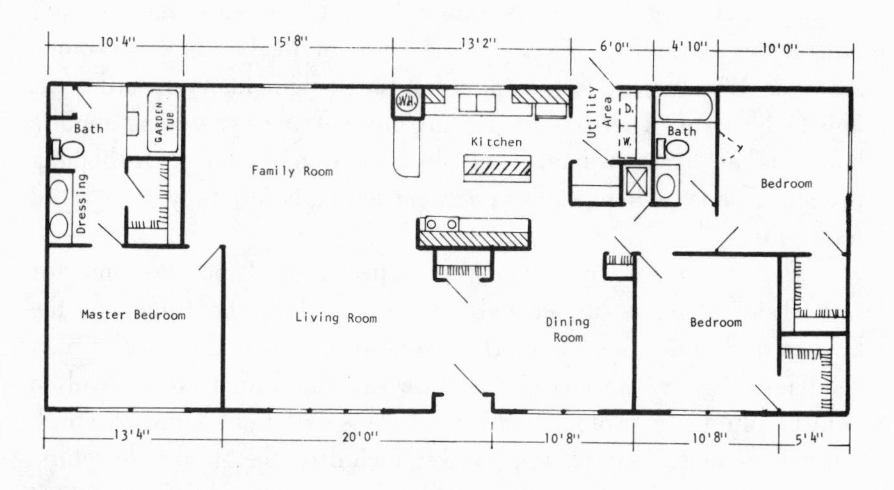

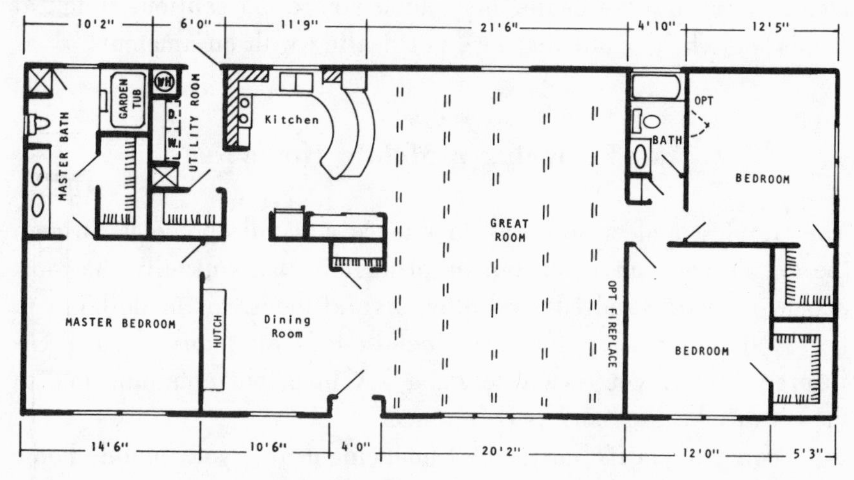

Surprising variety of mobile home plans is shown—starting with a basic 14-foot wide section up to 70 feet or so long. Some manufacturers also provide optional sections with one or more rooms that can be added at right angles to the basic house. Redman Homes.

because used mobiles often cost a lot less than new ones. A used regular house, on the other hand, generally costs almost as much as a new house, even though the condition of the used house has depreciated. That's largely because the value of the land under the house has increased.

Special bargains are available when repossessed mobiles are put on the market. That's when the bank or dealer takes over one that was largely paid off, and is not stuck with many installments due to be paid. There is a growing market in repossessed mobile homes taken back by banks and dealers. Repairs and refurbishing are often made to them, such as new carpeting, before they are offered for resale.

Here are a few tips when you inspect a used mobile home for sale. Take along a rubber ball and place it in the center of the kitchen and bathroom floors. If it rolls to a corner, the mobile may need leveling, or the chassis may be sagging which could lead to painful plumbing problems. Also take a small light lamp to check all wall sockets. Test *all* appliances, including the smoke detectors. Don't worry that the dealer may think you're too cautious. Look at it this way: he'll know that he's not dealing with an amateur!

Financing a Mobile Home

The simplest and easiest way to finance a mobile home is through the dealer you buy from, but unfortunately that's usually the most expensive, too. Hundreds and often several thousands of dollars can be saved if you shop for financing. Best of all loans is an FHA-insured one; if you're a veteran, a VA loan, but obtaining one of these can take time and perseverance.

You can qualify for the best financing deal if your mobile home is installed over a foundation, like a regular house. Then you often can get a regular house mortgage loan at the same down payment, 20-years-or-more repayment schedule and the same regular interest rate that is used for regular house mortgage loans. A regular mortgage loan is unquestionably the kind of loan to get for a mobile home anchored to a regular foundation. It's obtained from many of the same banks, savings and loan associations, and other lenders who give regular house mortgage loans. The mobile home dealer you buy from should know who they are in your area.

Financing a conventional mobile home tied to a mobile pad and not anchored to a foundation costs more because it requires the same kind of installment loan used for financing cars. The down payment required ranges from nothing down for a veteran buying with a VA-guaranteed loan, up to about 25 percent down. The most common down payments fall in the 10 to 20 percent range.

The monthly payments depend of course on three variables: the price of the mobile you're buying, thus the size of the money egg to be repaid; the interest rate charged; and the length of time required to repay it. The longer the repayment period, the more you can stretch out your monthly payments and the lower each monthly payment. But you pay for that privilege by having to pay more interest over the years. The repayment period can range from a short seven years up to twenty, though this (and other financing facts given here) can change from year to year from 13 to 15 percent, more or less.

If you can obtain an FHA- or VA-guaranteed loan, the interest rate can be less and over the years it can add up to quite a few dollars saved. FHA and VA loans are also obtained from private banks and other lenders, and they can offer other advantages such as a lower down payment. FHA and VA loans are guaranteed by the government so there's less risk to the lender, and therefore lower cost (reduced interest rate) for the borrower (the mobile-home buyer). Unfortunately, obtaining an FHA or VA loan may involve time-consuming red tape, which can be a pain. But it's a small price to pay for the dollar savings they offer.

No matter what kind of financing is sought, the importance of shopping for a good loan at the lowest cost cannot be overstated. On the whole, you will generally save by obtaining your own loan direct from a lender, rather than taking the easy road and financing the deal with your friendly smiling mobile-home dealer who will be delighted to handle all the details for you. It's convenient, all right, but it can also cost you extra money. After all, the dealer must do extra work so why shouldn't he be paid extra?

9. HOW TO BUY A THOROUGHBRED FACTORY-MADE HOUSE: JUDGING DESIGN

Most of us can spot a handsome woman at a glance but for unaccountable reasons we are taste-blind when it comes to houses (and cars). Having money makes little difference. Many houses in the richest suburbs have no more style than a neon sign.

In short, they lack good design. Don't shrug off the word "design." It may sound highfalutin but it goes deeper than what you see when you first look at a house.

Besides being appealing to the eyes, a well-designed house—factory-made or not—fits well on its site, catching loads of bright warm sun in winter and thus lower winter heating bills. The same house also knows how to brush off the hot sun in summer and thus, lower cooling bills; or, without airconditioning, it's naturally cooler in summer.

Its interior floor plan and room designs make it easy, convenient, and sheer pleasure to live in. It's built of good-quality construction, which means it's easy to keep clean and maintain and its cost of upkeep is low. This, of course, is more likely with a factory-made house than with a stickbuilt one. The payoff is that a truly well-designed house maintains really high resale value over the years. Though many houses have increased in value in recent years, the value of some have risen much more than others. Witness, for example, the high value put today on the best Colonial, Federal, and other traditional houses of yesteryear. And, perhaps most of all, the highest prices are paid today for the best contemporary twentieth-century houses designed by top architects like Frank Lloyd Wright.

The same design principles that make such houses good apply to any factory house that you may buy, as well as any other house. Here and in the following chapters are the basic things to know when you shop for a manufactured house: good style and appearance, house-to-site relationship, good floor plan and room design, and good-quality construction.*

The Essence of Handsome Looks

The key thing is that a good house has form. All of its parts fit together in the right proportion. They form a coherent whole. There are no jarring notes. The facade, for example, is clean and simple. It is not broken up with a banana-split mishmash of different materials, plus redundant shutters on the windows. Those are worse than warts.

Here are other points to avoid when you look at factory house brochures and photographs. The walls should be neat and straight and not jut out here and there for no reason. The roof should sweep across the length and breadth of a house also with no unnecessary broken lines and changes. Breaking up the sweep of the walls and roof of a house with jogs and turns is supposed to add variety and interest but usually it adds nothing but chaos.

Notice how well-designed houses—those designed by architects and shown in magazines, for example—are such that the tops and bottoms of the doors and windows almost always line up across the facade of a house. When the opposite occurs, with doors and windows located at different levels as if they were thrown into the wall

* Some of the material that follows is from the author's book, *How to Avoid the Ten Biggest Home-Buying Traps*, copyright © 1969 by A. M. Watkins.

willy-nilly, the house looks as scrambled as eggs. It's out of scale, and overall it seems smaller than it actually is. That's one of the many subtle design blemishes that detracts from resale value.

Watch out for gingerbread, which may be fine in Hansel and Gretel, but is usually the kiss of death for a house. Adding a lot of unnecessary clutter and doodads to a house, like iron grilles, and fancy woodwork, as well as unnecessary shutters, is like piling cheap jewelry over a flashy dress. It's out of place. Now shutters and grillwork may be in place in an authentic Georgian Victorian house, but that's something else again.

An excellent little discourse on good house design and appearance is given by architects John L. Schmidt, Walter H. Lewis, and Harold Bennett Olin in their book, *Construction Lending Guide: A Handbook of Homebuilding Design and Construction,* written for

This house not only shows handsome results of good design, but is further proof of the good architecture in manufactured houses. Deck House.

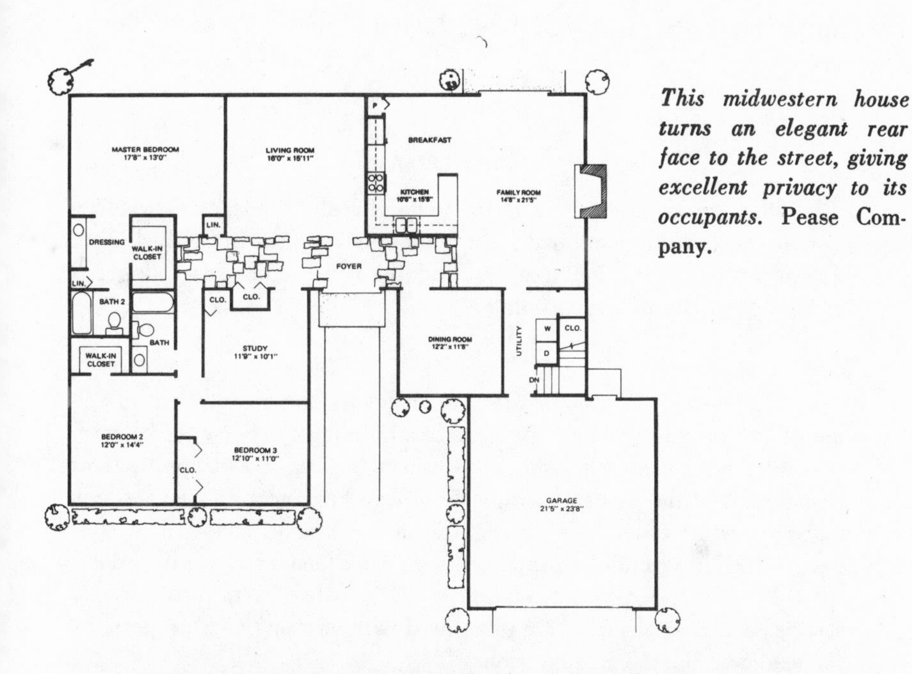

This midwestern house turns an elegant rear face to the street, giving excellent privacy to its occupants. Pease Company.

the United States Savings and Loan League. Here are some things they say: *

EXTERIOR APPEARANCE

Pleasing appearance is no accident. Careful study and organization are necessary to anticipate how a house will appear in finished form. The appreciation of a well-designed building is based on recognition and evaluation of the following points.

PROPER PROPORTIONS

There are basic combinations of shape and mass that result in balanced building proportions. Portions of a building can be out of balance, just as a scale can be tilted. In traditional styling, the proportions of structures and the elements within the design have been refined through many years of study. Roof slopes, overhangs, window shapes, and sizes have all been carefully considered to fit with one another. In today's typical house design (in attempting to achieve the charm of traditional styling) familiar elements are often used without exercising proper care for achieving pleasing proportions.

VISUAL ORGANIZATION

A house is a complex arrangement of parts and pieces. Success in exterior design rests in large part on the visual continuity of these elements, which should be related in shape, form and arrangement. Visual organization is the assembling of the parts and pieces with these relationships in mind.

MATERIAL USAGE, TEXTURES

Materials should be selected for uses appropriate to their capabilities. The elements of a house should be built of materials capable of performing satisfactorily, both initially and over the years.

The visual response aroused by various materials differs greatly. For example, the smooth, "cold" flatness of porcelain-enamel steel panels affects a viewer very differently from a rough, nubby stone wall of rich "warmth." The texture of materials is extremely important in the design of houses and in the selection of materials to be used.

In general, the number of textures selected should be held to a minimum; one type of masonry, one type of wood, or one siding texture, and one neutral "panel" surface per house. Contrasts can be used very

Two-story open planning makes the inside of this house glamorously bright and large. The big windows can let buckets of sunshine in to help heat the house in winter, providing they face south. American Barn/Habitat.

L-shaped deck serves not only as a natural outdoor extension of the living area of this house, it also adds greatly to the size and attractiveness of the house. Acorn Structures.

successfully, just as a man can be well-dressed with a coat and trousers of different texture. The well-dressed man, however, and the well-dressed house do not wear many varying materials at the same time.

SCALE

Scale is the relationship of design elements to the human being. One's visual sense depends in great part on scale relationships in judging distances, sizes and proportions. The size of a door in a house facade can be "in scale," that is, proportioned agreeably to the human being and to the rest of the house; or it can be "out of scale," that is, not properly related to the human figure, overpowering or diminutive in the entire design.

SIMPLICITY AND RESTRAINT

To simplify is to refine. In housing, the simplest, most basic designs are the hardest to achieve but simple visual elements are the most pleasing. The attempt to make a house appear more expensive by cheap imitation of expensive items can be disastrous.

Scrollwork and carpentry bric-a-brac can result in design chaos. Such an approach is often based on using many unrelated elements in an effort to develop curb appeal. A storm door with a pelican scroll, a checkerboard garage door, slanted posts at the entranceway, diamond-shaped windows and other parts and pieces having no design continuity detract from appearance and result in a modest house looking cheaper instead of costlier.

The most successful approach is one of restraint. Generally, the simply stated house design is the handsomest. Restraint and sophistication go hand in hand. An opulence and overdecoration are more often than not annoying and disturbing. The simplest designs are the most agreeable in the long run.

COLOR

Color is a highly potent factor in the design of a group of houses taken as a whole. In this respect, color coordination is a major way to strengthen that pleasing individuality from house to house, the objective of good subdivision design. Color is actually so powerful that it can make a cracker-box house look attractive or a generally well-designed house look repulsive. It all depends on the skill with which color is used. When a large group of houses is involved, the overall effect of color is a major concern. The cluttered, disorganized look of the average subdivision of low-to-moderate-priced houses (or even more expensive ones) is due largely to lack of color coordination. Clashing roofs, anemic body

Here's how natural wood siding is used effectively in contemporary houses of one- and two-story designs from the same maker. Notice how the lower level of the one-story model, ordinarily a limited-use basement, has been opened up with sliding glass windows and door to take advantage of the sloping land. Northern Homes.

colors, misplaced accent colors—all these result from the short-sighted practice of giving the buyer too much latitude in selecting exterior color.

THE ALL-AROUND LOOK

Unfortunately, most houses are designed like a Hollywood set. Some concern is evidenced over the appearance of the street facade, but often the side and rear elevation are totally neglected. A well-designed house, like a piece of sculpture, should be handsome when viewed from any vantage point.

A FEW WORDS ABOUT STYLE

A well-known architect, Alden Dow, FAIA, puts it simply: "Style is a result, it can never be an objective. When style itself becomes the objective, nothing results but a copy."

What determines style? The shape and character, or style, of a house should be determined by the plan, the site, methods and materials of construction, and by the budget. A particular set of circumstances, worked upon by the design process, logically will lead to a particular set of building shapes or appearance. It might be correct to say that the well-designed house is styleless, since no forcing of the solution has been made by adapting it to the framework of a "traditional" scheme. Traditional styles are, of course, in predominance and undoubtedly will remain so for many years. But there are other styles to consider.

Here are four terms often discussed in describing other than traditional architecture: modern, modernistic, contemporary, and futuristic.

A modern house is one built of up-to-date materials: it has most of the current electrical and mechanical gadgetry in place and may exhibit itself as a simplified expression of any of the whole bag of traditional styles. It is harmless in design—not bad perhaps—but not great architecture.

A modernistic house is a poorly designed "jazzy" modern house. Often it is an attempt at being unusual, generally is designed by a contractor or individual and is usually a collection of pieces (perhaps a flat roof or a butterfly roof, round windows, slanted posts, big "picture-windows") with no integration of parts into a carefully studied design.

A contemporary house generally is one done by an architect and grows in its design from the consideration of beauty, function and site. The market confuses good contemporary design with modernistic and fails to perceive the great difference between the two. Contemporary design is not faddish, or subject to being "in" today, "out" tomorrow. On the contrary, it's extremely rare, especially in lower-cost houses.

A futuristic house is somewhat experimental in nature. New

products or methods may be tried or tested, and generally the futuristic house attempts to present an image of tomorrow's house. It may be, and usually is, a properly conceived design, but its importance on the market is inconsequential.

Poorly designed modernistic houses invariably will be penalized by the market, but an honestly conceived contemporary home, large or small, is of lasting value. Little research is necessary to recognize that a fine piece of architecture appreciates in value and appeal. Certainly the demand far exceeds the small supply.

10.
HOW TO TEST THE FLOOR PLAN AND OTHER ROOMS

It's like the man who thinks that meat and potatoes is the only meal to dine on. He doesn't know any better.

Similarly, many a person doesn't know that a floor plan is more than a few rooms connected in one way or another. Afterwards he wonders why living in a house with a sick plan can drive a person up a mountain.

That's the difference between a factory house with a poor plan and one with a good plan, including good interior room design and things like good "zoning." These can make an enormous difference in the pleasure and sheer enjoyment that you get—or don't get—from a house. Here is what to know about each when you shop.

The key to a good floor plan is that it permits good people circulation from room to room and also in and out of a house. Like highway traffic, you want to avoid snags, tie-ups, and jams. Apply the following tests to each plan. But also be flexible, especially if you're buying a mobile home or other compact house where the design and the floor plan in particular sometimes must be compromised a bit for economy reasons. On the other hand, a good number of manufacturers will vary their standard floor plan for you on request and often with little or no extra charge.

1. *The main entrance,* the front door, should funnel people, mostly visitors, directly to the living room. An entrance foyer is highly recommended for receiving guests. A coat closet nearby is virtually essential.

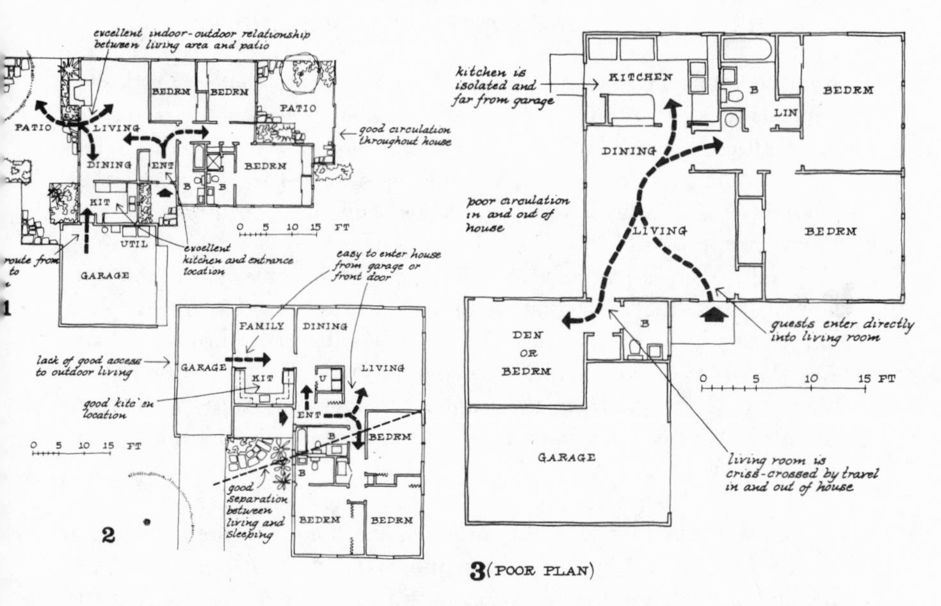

Floor plans show important features to have in your house and typical mistakes to avoid. The Building Institute.

The main entrance should be quickly and easily accessible from the driveway and street in front. It should also be quickly accessible from the rooms inside, where you are likely to be when the doorbell rings, and especially from the kitchen. The kitchen-to-front-door route is one of the most frequently used paths. A foyer is also important as a buffer or transfer chamber to keep howling winds, snow, and rain from blowing into the heart of a house every time someone comes in the front door.

2. *A separate family entrance,* ordinarily a back or side door, should lead directly into the kitchen area. This is the door most used by a family itself, especially children. A proper location is important to permit swift unloading of groceries, for example. It should also be located so that children can travel in and out easily and can quickly get to where they're going inside the house (such as a nearby bathroom). The route from the car to this entrance should be sheltered from rain and from snow and ice in a cold climate. But the route from it through the kitchen should not run smack through the kitchen working area.

3. *The living room* should have a dead-end location for the reasons noted earlier plus a few others. It should not be a main route for everybody's travel around the house. Only then can you entertain guests in peace or just read or watch television without being continually disturbed by others walking through. Sometimes one end wall of the living

room serves as a traffic lane; in effect, it's a hallway. That's all right if it happens to work in a particular house. Sometimes, though, a screen or half-wall may be necessary between it and the heart of the living room.

4. *The room arrangement* should be designed so that you can go from any room in the house to any other without going through a third room except possibly the dining room. Direct access to a bathroom from any room is particularly important. Yet in some houses this rule is violated to the consternation of its occupants, and there are even houses in which access to a bedroom is possible only through a bathroom! This is unpardonable, no matter what kind of factory house you may buy. If the bathroom is occupied, a person is trapped in the bedroom like a prisoner, unable to get out (unless he doesn't mind going out a window).

5. *The kitchen* should have a central location. It should not be located way out in a left-field corner of the house. In the kitchen one should be close to the front door, should be able to oversee children playing in the family room, say, or outside, and should be able to get to the dining room, the living room, or the terrace without long hikes back and forth. That's particularly important when entertaining.

6. *The main travel routes* between the house and the outdoor living areas—patio, terrace, or porch—should be short and direct. Can guests as well as family members go in and out easily? Is the outdoor area that you use most during pleasant weather easily accessible from the house? If it is, it will be used often. If not, you'll find it neglected and your family involuntarily depriving itself of outdoor-living benefits.

Those are the characteristics of a good floor plan. They're not always easy to get, which is why there are so many garbled plans (though this is less true among factory houses than local builder houses not designed by architects). In fact, making a really efficient floor plan is one of the toughest things to do in the design of a house. The importance of some of the individual tests will, of course, vary from family to family, depending on your own living habits and the activities you consider most important.

If you like to spend a lot of time outside the house, for example, you may place high importance on efficient access to the outdoors. If you entertain often, the design and location of the dining room and living room for entertaining are vital for successful social events. The way you live, therefore, should be taken sharply into account so that your house will simplify daily living activities and make life a lot more pleasant.

Interior Zoning

Interior zoning is concerned with the logical arrangements of the rooms inside the house. Ideally, every house should have three clear-cut zones to accommodate the three main kinds of activities: living, sleeping, and working. The living zone embraces the living, dining, and family rooms, in which you engage in most activities other than working and sleeping. The work zone embraces the kitchen, laundry, and perhaps a workshop, where obvious, if not unavoidable, work of one kind or another goes on. The sleeping zone embraces the bedrooms. Each zone should be separate from the other. The two-story house provides natural zoning between the bedrooms upstairs and the other rooms downstairs with natural benefits.

Regardless of the kind of house, a buffer wall or other such separation is essential between the bedrooms and the other two zones, if for no other reason than to permit you to entertain guests without disturbing children at study or in bed at night. The kitchen and work zone should be separate from the living area. Can dishes be left stacked and unwashed there without being seen by guests in the living room? Can laundry be left unfinished but out of view when visitors call unexpectedly (or even when they are expected)? The answer should be Yes. This type of question will tell you if a house plan has good zoning.

The Kitchen

The kitchen clearly deserves top-priority attention. A homemaker usually spends more work time there than in any other single room. It also represents a very high cost part of a house (because of its heavy-equipment concentration). Its cost can run half as much again as the average square-foot cost of a typical house. The kitchen more

This inviting foyer is no accident. It is well planned.

than any other room also tends to influence the resale value of your house. It is clearly important, and its design elements should be explored in some detail.

A central location is the first requisite, as noted earlier. The kitchen also rates a good exposure in relation to the sun. The same principles apply as those given later in this book, chapter 13, for overall orientation of a house. Best kitchen exposure is one with windows on the southeast; next best is south. That will let most bright light and sunshine flood in during the day for at least the eight months from September to April. It will mean a bright, airy kitchen most of the time when you want it, yet the same exposure is easy to shade on hot summer days. With a factory house, this of course means that the house be tailored for its lot, or vice versa.

A kitchen facing east gets sun in the morning, but that's about all. One facing west gets little or no sun except in the afternoon, and in summer it will get hit with particularly hot afternoon sun. A northern exposure is darkest and gloomiest of all, receiving the least sun and light the year around. The same exposure principles also apply to a dining room or other area that is used for most meals.

Kitchen Work Triangle

The heart of a kitchen is its "work triangle," the arrangement of the refrigerator, sink, and range in relation to each other. The entire process of efficiently preparing and cooking foods hinges on a good work triangle. From the refrigerator to sink to range should form a triangle with a total perimeter of at least twelve to fifteen feet, but no more than twenty-two feet, according to research at Cornell University's renowned kitchen laboratory. The appliances should be in that order to conform with the natural sequence of cooking.

Plenty of countertop space around the triangle is also a must. The University of Illinois Small Homes Council recommends these minimum standards: at least 4½ feet of countertop length on the open-door side of the refrigerator between the refrigerator and the sink; 3½ to 4 feet of countertop length between the sink and range; and at least 2 feet of countertop on the other side of the range. That adds up to a minimum of at least 10 feet of countertop length in all. An additional 2 feet of countertop is desirable, if not essential, at or near the range as a last-step serving center, where food is put on plates before being taken to the table. If either the refrigerator, the sink, or the range is separate from the other triangle centers—on a separate wall, for example—extra countertop space should be placed at its side, in addition to the minimum standards just given.

Make sure the refrigerator door opens the right way—toward the counter between the refrigerator and sink—so food can be conveniently unloaded where you want it. The wrong-door refrigerator is a common flaw. A separate wall oven can go almost anywhere. Once it's loaded, it can usually be turned on without demanding attention until the bell rings. A location near the range is not essential, but because of its heat, it ordinarily should not be put flush next to the refrigerator.

Any ample kitchen core and work triangle usually require a space at least 8 feet by 12 feet (96 square feet). That's the minimum to look for. With a dishwasher and separate oven, more space is needed. This should be an exclusive self-contained part of the kitchen out of the way of the main traffic routes used by people passing through the kitchen from one part of the house to another.

The kitchen, though usually fairly well-designed, or better, often can be modified, if necessary, to meet any of these top standards.

Kitchen Storage

According to studies at the University of Illinois, a minimum of at least 8½ running feet of wall cabinets and/or storage shelves is recommended for the kitchen. Another rule calls for at least 20 square feet of interior storage space under the countertop plus at least 10 square feet in wall cabinets. The proper cabinets and shelves should be located where they can house the particular items needed in each part of the kitchen. For example, storage for dishes and pots and pans near the sink and range; storage for working knives, bread box, flour, and other staples near the sink-refrigerator center, and so on.

Because kitchen cabinets can run into big money, consider open shelves for certain items. They can be a lot cheaper. If not enough cabinets or shelves are in a house, space should be available against walls or under the countertop to add what you'll need. The quality of cabinets is also important. You'll certainly want attractive cabinets with a rugged, hard finish that is easy to keep clean and will stay attractive over the years. The drawers should roll in and out easily, which calls for nylon rollers; try them and see. The cabinet hardware and latches should be of good quality.

Fixtures and Appliances

The best kitchen sinks are made of either stainless steel or enameled cast iron. Both are easy to keep clean and will retain their good looks over the years. There are also porcelain-on-steel sinks, which may look like cast iron; they are hard to keep clean, chip easily, and lose their gloss quickly. Sometimes the material of which it is made will be noted right on it; other times you must ask about it. You'll probably want a single-lever "one-armed" faucet rather than the old-style double-handle type. (More on faucets in chapter 11.)

Appliances are largely a matter of personal preference. Sometimes when buying a factory house the appliances are optional, and you may bring along your own. You should know, however, which ones do or do not come with it. Those that are included should be noted in the sales contract to avoid a common misunderstanding that occurs when people buy houses.

Gas or Electric Range?

This is particularly important because electric cooking costs about two to four times more than cooking with gas. The exact extra price paid for electricity depends on the local cost for each in your area. Even though gas rates will probably rise faster in the future than electric rates, gas should remain significantly cheaper for a long time. Gas costs will have to rise by more than 200 to 400 percent, which is quite a bit, before they reach the cost level of electric cooking.

The dollar cost of cooking with gas comes to about $5 to $10 a month versus $20 to $30 a month for electricity. Calls to your local utilities should get you specific figures on the comparative cost of gas and electric cooking for your house.

In general, you can save about $10 to $20 a month with a gas range, so gas is obviously the more economical choice. An electric range for the kitchen is recommended only when you have a strong personal preference for electricity that offsets paying more—as much as $200 or so more a year—to cook with electricity.

Gas is cheapest of all if it is also used for the house furnace and water heater. Then your gas cost will fall to the lowest rate. That's a result of higher use; the more you use, the lower the rate. Similar reduced rates on a step-down schedule also apply for electricity; the more used each month, the lower its unit cost. That is often mentioned when electric companies promote the all-electric house.

But this can be misleading even if you must accept electric heat (because gas or oil heat is unavailable). Even then it can still pay to use gas for the kitchen range and water heater. That's possible even in areas where gas is unavailable for house heating but still can be used for other house uses.

Your total energy bill each month for electric heat and gas for cooking, water heater and other permissible gas use will often be lower than the cost of electricity for all your energy. The exact savings vary according to local energy rates, which you must figure yourself according to the local rates for each. More on this is given in chapter 12 on energy.

Final Checks for the Kitchen

Try to visualize the overall kitchen. It should be large enough to hold the table size required by your family, or an adequate dining area should be nearby. Some people also like space for a work desk and perhaps a sewing table. The laundry also should be nearby, or you may desire space in one part of the kitchen for a washer, dryer, and ironing board. A laundry located on the second floor of a two-story house can be the best of all. It not only saves stair-climbing, but puts the equipment at the largest source of the wash.

Are there electric outlets spaced behind the countertop for convenient use of small appliances? If not, how will you operate a mixer, a blender, and an electric frying pan as well as a toaster and a coffee maker? You'll want good lighting from above and lights that shed illumination over the full length of the countertop. Finally, there's the need for good ventilation to keep the kitchen (as well as the rest of the house) free of cooking fumes and odors. A built-in exhaust fan is the least required. It should be located in the wall directly behind and above the range or in the ceiling directly over it. If located elsewhere, its exhaust efficiency will be low. A large range hood with built-in fan is even better. It should exhaust outdoors through a duct. There are also "ventless" hoods equipped only with filters, with the hot air drawn through the filter and then spilled back into the kitchen; they do not cool a kitchen.

Other Rooms

Windows make an enormous difference. They are the biggest single ingredient for making a room cheerful, large, and pleasant. They turn a small space into a virtual ballroom, if necessary.

So look for ample window glass to let in light, air and sunshine, the main reasons for windows. But obviously they should not let in a lot of cold in winter and hot sun heat in summer. You can have your cake and eat it, too, by careful design and location exposure of the windows. In general, a southern exposure for your windows is best of all, east or west next best, and north is worst of all.

At the same time nearly every room needs enough unbroken wall area for easy furniture placement. Rooms should be large enough to accept your furniture. Good heating, enough electric outlets, and good lighting are also points to check. As for size and location, here are some minimum standards from the University of Illinois Small Homes Council. (Don't be put off by the word "small" in their name.)

- A living room at least 12′ x 20′, with at least 10 to 12 feet of unbroken wall for a couch. Remember, the living room should not be a major highway for traffic through the house, and the front door of the house should not open directly into it.
- A family room of at least 12′ x 16′, though 12′ x 20′ is better. It should be on the same level as the kitchen and near, if not adjacent to, the kitchen.
- Bedrooms at least 9′ x 11½′, with at least 4 square feet of closet space per person. The bedrooms should have privacy from the rest of the house, and the master bedroom in particular should have built-in privacy from the children's bedrooms. You'll notice that many notable custom houses published in architectural magazines will have the master bedroom located apart from other bedrooms, if not by itself on the other side of the house from children's rooms.

That's the result of architects designing individual houses to fit the needs of individual families. It's not easy to do in a modest-priced house, and it's not too often found in factory-made houses. But nonetheless, it sometimes can be had by modifying a factory-house plan. Think about it. It can make an excellent feature that you will enjoy.

I've mentioned that a 12-foot dimension for one side of any room of a house, including living and family room, will provide a satisfactory room. That's also generously ample for a good bedroom. It will conform with the minimum standards for a good room.

Common House Traps

Here are 22 common little flaws in factory-made as well as other houses. They include a few mentioned elsewhere but repeated for emphasis because they can be a big pain to live with. They belong in no house. This list also shows how good design embraces the important little things, as well as the big, and that good design involves more than style and good looks. The flaws are:

- No separate entranceway or foyer to receive visitors.
- No opening in the front door, or no window or glass outlook alongside that lets you see who's at the door.
- No roof overhang or similar protection over the front door for shelter from rainy weather.
- No direct access route from the driveway to the kitchen.
- No direct route from outdoors to bathroom so children can come in and out with minimum of bother and mud-tracking.
- Gas, electric, and water meters inside the house or in the garage or basement, rather than outside. Outside meters do away with the need to let meter men in every month.
- Fishbowl picture window in the front of the house, exposing you to every passerby.
- The nightmare driveway that opens out on a blind curve so you cannot see oncoming traffic when backing out. A driveway that slopes up to the street is almost as bad, especially for trapping you hopelessly on a winter morning when your car won't start.
- Isolated garage or carport with no direct or protected access from car to house.
- Accident-inviting doors that open toward the basement stairs.
- Cut-up rooms with windows haphazardly located. Sometimes too many doors make it impossible to arrange furniture.
- Windows in children's rooms that are too low for safety, too high to see out of, and/or too small or difficult to get out of in case of fire.
- A hard-to-open window, usually the double-hung type, over the

kitchen sink. An easily cranked casement window is usually best here; a sliding window second best.

• A window over the bathroom tub. This generally causes cold drafts as well as rotted windowsills as a result of condensation.

• Stage-front bathrooms placed squarely in view of a space like the living room or smack in view at the top of the stairway. Ideally, one should be able to go from any bedroom to the bathroom without being seen from another part of the house.

• Only one bathroom, especially tough on you in a two-story or split-level house.

• No light switches at every room entrance and exit.

• No light or electrical outlet on a porch, patio, or terrace.

• No outside light to light up the front path to and from the house.

• Noisy light switches that go on and off like a pistol shot. Silently operating switches cost only a little more, and no new house can be called modern without them today.

• Child-trap closets that can't be opened from inside.

• Small economy-size closets that are hardly big enough for half your wardrobe. Also watch out for narrow closet doors that keep half of the closet out of reach without a fishing pole, basket-ball-player shelves too high for a person of normal height, and clothes poles so low that dresses and trousers cannot hang without hitting the floor.

11.

GETTING A GOOD BATHROOM, THE HIGHEST-COST ROOM

The bathroom rates a chapter of its own not just because it's the highest-cost room in a house, but also because it must withstand harder use and abuse than any other part of the house. It therefore pays to be sure you get good ones in the factory house you buy.

That means a bathroom that's tough and easy to keep clean and attractive, requires a minimum of service and maintenance over the years, and will not start looking distressingly rundown and shabby in a distressingly short time. Knowing how to get a good bathroom is particularly important with a factory house because the bathroom fixtures and other parts often must be provided and installed by the buyer since a number of manufacturers do not include this part of the house with their house packages.

The location of each bathroom is also important. At least one should be near the bedrooms. If it's the only one in the house, it obviously should be convenient to other rooms as well.

Two or three bathrooms are usually essential for a family, especially in a large house. If an extra bathroom is wanted in a new house, order it when the house is ordered rather than after you move in. It can be installed for much less cost at that time than when the house is completed. If that requires more extra money than you can afford, try to allow for the future location of the additional bathroom you'll need later. Have the main plumbing pipes supplied to it and capped for future use. The stitch-in-time cost to do this can be small and the savings later quite large when you complete the installation.

A big selling feature in houses is the private bathroom for the master bedroom, but this is not always good. In a two-bathroom house it might be better to have a house with the grown-ups' bathroom outside the master bedroom, where it is accessible to guests. Otherwise guests may be restricted to using the children's bathroom, which is often an embarrassing mess. Free the master bathroom, and the second bath can be given over completely to the children with no worry about the inevitable disorder found in it. Most manufacturers will change the location of a master bathroom on request. This is unnecessary, of course, if a third bath or "powder room" is available for friends and visitors.

A full bathroom may be as small as 5′ x 7′, and a half-bath (powder room) as small as 24″ x 40″ or so. However, the key to adequate bathroom size is not the dimensions necessarily, but the number of people who are likely to be using it, particularly during the hectic morning rush hour. The more baths, the smaller the load on each and the less the need for large baths. You can usually tell by sight if a bathroom is large enough to handle two or three children or two adults at once.

But extra bathrooms are not always necessary. Two lavatories, or washbowls, for example, may save you the expense of an additional bathroom. Two of them side by side can provide double-duty and help to solve the morning rush-hour problem. Sometimes a bathroom can be equipped with two toilet compartments, which also may be adequate and less costly than an extra bath. These are good ideas for families with children.

Compartment Plan

A whole bathroom for a large family can be compartmentalized, an idea heavily promoted by fixture manufacturers, though not found in many manufactured houses. If desired, you could ask about getting this in a factory house you buy. Toilets, shower, and sometimes the tub are fenced off by partitions which enable from two to three people to use the bathroom simultaneously with privacy. Partial partitions, dividers, or folding doors may be used. If two toilets are used, each has a private enclosure like a public bathroom. Two washbowls are mandatory, a second tub or shower stall is optional. A compartment bath is equivalent to having two or three bathrooms, but plumbing and construction costs are lower as a result of grouping all the fixtures together within one wall enclosure. A space of at least 8′ x 10½′ is required to start with, 10′ x 12′ is better.

The merits of the compartment bath, however, are open to question. It has received much publicity, because large fixture manufacturers landed on it as a good way to sell more fixtures—two instead of one at a crack. Merely partitioning off the toilet in a large bath coupled with double washbowls often can serve a family as well at less expense. Or you may do better with a separate second bathroom elsewhere, especially in a large house.

Choosing Good Bathroom Fixtures

The old traditional fixtures—bathroom sink, toilet, and tub—are made of china, steel, or cast iron. The newcomers are made of rigid plastic which are increasingly used especially in manufactured houses. Here I'll consider the traditional materials first, though they're not always installed in manufactured houses. But if available and desirable, it's important to get good-quality ones. Rigid plastic fixtures can be just as good, and sometimes you have no choice. You must take them because they're standard equipment. More about them is given later.

Traditional washbowls, toilets, and tubs come in three quality grades. Like many products in the Sears, Roebuck catalog, they're called "good," "better," and "best." But that's namby-pamby. What's called "good" is in many cases cheap, low quality that shouldn't be given to dogs.

In order of increasing quality and durability, washbowls are made of enameled steel, the bottom-of-the-line; enameled cast iron, the middle; and vitreous china, the top-of-the-line. A steel bowl is least desirable because it is difficult, if not impossible, to fuse a durable enamel finish on steel. In time the finish pops off.

The finish on both steel and cast-iron bowls is porcelain enamel, a form of melted glass. It adheres to cast iron much better than to steel, so it's silly to accept steel. An enameled cast-iron bowl is far more chip-resistant and has a permanent finish. It's especially recommended in a child's bathroom where it can take tough knocks without showing them (which is why it's often specified for schools and hospitals).

The vitreous-china bowl, the top-of-the-line, is not necessarily as chip-proof as enameled cast iron, but it's still very tough indeed. It is top quality because of its gleaming finish and handsome looks. It's also more versatile than cast iron because it can be cast into a greater variety of shapes and sizes and only costs a few dollars more than cast iron. A china bowl is usually the best choice and enameled iron next best.

Whichever you choose, don't be cursed with a midget bowl. Some are so small that merely washing your hands will bruise your knuckles. They're also too small for washing your hair or bathing a new baby. Once hooked up, you're stuck with it. Yet the price difference between a luxuriously large 20″ x 24″ bowl and the small but commonly sold 17″ x 19″ model is no more than the price of a few steaks.

The Bathtubs

Bathtubs are made of enameled steel or enameled cast iron. They're not made of vitreous china because china can't be molded in large sizes except at prohibitive cost. The enameled steel tub is the cheaper kind, but it's not all that bad, provided it's not subjected to rough use. It can be manhandled into place and installed by one person, an advantage in do-it-yourself installation. Cast-iron tubs require two men for installation. A cast-iron tub is usually a better buy because of its durability and superior resistance to chipping. It costs about $100 to $150 more than a steel tub.

When a spanking new tub is first used people often discover that it's too shallow or small, or both. That's a typical tub—30 inches wide, 14 inches deep and 4½ or 5 feet long. So specify a larger size

—at least 32 inches wide, 16 inches deep, and 5 feet long. That increase may seem small, but it makes a big difference in bathing comfort, including bathing children. It also will mean less splashed water on the floor. Also be sure to get a slip-resistant tub bottom.

To cut costs, many fixture makers have cut the depth of their tubs to as low as 12 inches, which is low. It's hardly off the floor. Shallow depth is usually what you will get unless you specify a tub at least 16 inches deep. This is practically essential if you like tub baths and want to avoid wild splashing over the floor with children.

A square tub shape can be interesting if it is large enough and has a seat at one or both ends. People need one at least 4 feet square but often get saddled with a tiny thing hardly larger than a washbowl.

Bathtubs come in a rainbow range of colors to choose from. A colored tub generally costs 5 to 10 percent more than a white tub. Tubs are also available in a variety of models designed especially for the hedonistic person who craves luxury bathing: tubs with body-contoured shapes, reclining-back designs, beveled headrests, and for the true Roman bather, ultradeluxe poollike tubs.

Toilets

There are three basic toilets, according to government standards:

The washdown which is the aboriginal toilet, unchanged since the Spanish-American War. It's the bottom-of-the-line model. It needs frequent cleaning, is not sanitary, it's also prone to clogging and it's noisy. A washdown toilet is easy to spot because the water drains out the front of the bowl. Fortunately, some building codes prohibit the washdown. And it's being phased out by many manufacturers. But it's still available in the supply channels. Take care to avoid it.

The reverse trap, the middle-of-the-line model, flushes water out through a rear trap—thus its name. It is cleaner, more sanitary, and more efficient than the washdown because of greater water surface, deeper water seal, and large-diameter drain passage, which also makes it quieter. There are, however, marked differences among brands. Some have a greater-than-minimum water surface, larger-than-minimum drains, and a better flush mechanism. Their prices vary, depending on brand and quality of flush mechanism and by how much a model exceeds the minimum government standards.

Siphon jet, the quietest and most efficient toilet. Flush action is excellent, because water from the tank is discharged through jets located around the top rim of the bowl. It's also called a "quiet-flush" unit, and is identified by its one-piece construction and its low-slung shape.

There is also the handsome wall-mounted toilet which hangs entirely above the floor. It eliminates the big floor-cleaning problem encountered around a conventional toilet. This kind also can be comparatively quiet in operation if you get the right brand; some wall-hung brands, on the other hand, are noisy. The quieter kinds are identified by their low-slung tank and one-piece construction. A conventional high tank, separate from the bowl, is the tip-off to noisy operation, sometimes as loud and noisy as the cheapest washdown toilet. Check this in advance. A wall-hung toilet may require beefing up the wall, but this should not be a problem in a factory house.

Like the floor-mounted quiet-flush toilet, a wall-hung toilet is more expensive than standard toilets. But nearly every delighted owner of one or the other says it is well worth the expense, even if you have to work nights to pay for it.

Like cars, toilets come with optional features that you may want. These include an elongated bowl shape that is usually better and more sanitary than a round bowl, a self-ventilating bowl that is especially good for an interior bath, and the water-saver toilet which uses roughly one-third less water per flush than regular models. If you have high-cost water, this last can cut monthly water bills a good amount, since toilet flushes account for the major portion of most families' water use.

Those New Plastic Fixtures

The new rigid fiberglass and acrylic plastic lavatories, bathtubs, and shower stalls are being used in a growing number of factory houses, especially since they're light in weight and therefore mean reduced shipping and handling costs. They can be very good providing you get well-made ones. They are attractive, tough, and long lasting, since they're made of the same tough plastic used for boat hulls which must withstand greater exposure to water and hard knocks. Plastic units are also warm to the touch, a pleasure in winter, and are much lighter in weight than traditional metal and china fixtures. Their light weight can cut installation costs and be a heaven-sent boon for the person who builds his own house.

A one-piece plastic shower stall and tub can cut bathroom costs in a factory house not only because of their light weight, but also because they provide a hard, waterproof wall at less cost than tile or any other waterproof wall. A stall for tub or shower is made with integral floor and sidewalls (which does away with leaky grouted joints). The walls have a low-gloss finish which is impervious to alkalis, mold, fungus, and household solvents, but an abrasive detergent cleanser should not be used on them.

If you buy a factory house for which the plumbing is installed locally after delivery and plastic fixtures are to be used, be sure to get good-quality plastic ones. Many cheap, shoddy ones are available that are tempting to buy because of their low price. The plumber doing your bathroom installation may easily fall for these. Don't let yourself be a victim.

To protect yourself, specify a brand that meets the stiff design standards of the American National Standards Institute, or ANSI Z 124, 1-1974. A shower unit should meet the ANSI Z 124, 2-1967 standard. A similar standard is being developed for lavatories but until it's ready accept only a plastic lavatory made by a well-known company, which will replace it if something goes wrong. Another check is FHA acceptance of a manufacturer's unit. Ask about this. (Buying a product that comes with FHA acceptance is a good rule to use with many, though not necessarily all, housing products.)

Plastic fixtures tend to be priced higher than conventional fixtures but that's offset by lower-cost installation. A good plastic tub-wall unit, for example, may cost more than the lowest-cost enameled steel tub, but no more and maybe less than the same size top-of-the-line cast-iron model. And you don't have to spend more money to cover the surrounding walls.

Water Spigots, Faucets, and Shower Heads

Plumbers call these the fittings or the "trim." They are taken for granted by nearly everybody and we are then usually cursed with faucets that drip and trim that quickly tarnishes or rusts. Getting faucets that will operate long and well is important because faucets are among the hardest workhorses in houses. They must withstand harder and more frequent use than almost any other part of a house, with the possible exception of a swinging door in a house full of kids. So know a little about faucets. What follows also applies to faucets for the kitchen and anywhere else you have running water.

You can't count on getting good faucets and other trim simply by ordering a good bathroom bowl and tub. Fittings are ordered and shipped separately and are not always made by the manufacturer of the fixture. Like other parts of the bathroom, faucets are also made and sold in three quality grades: low, high, and luxury quality.

The bottom-of-the-line, cheapest faucets and trim are no good because they are usually made with quick-to-wear-out rubber washers and valves which is why they will drip like a head cold. They also need frequent regrinding of the valve seat.

Bottom-line faucets and trim (meaning water spouts, hand bars, other metal appurtenances) are also made of a relatively soft alloy —usually zinc with a thin top coat of chrome. They shine in the showroom, but once in use they tarnish as quickly as they drip. There are exceptions, but you can generally spot the cheapie faucets by their handles which have two or more spokes radiating out from the center, like a wheel. With a few exceptions, the better kinds have solid handles, with a ridged circumference for gripping.

Good faucets and trim, the middle-line kind, cost only a little more than the cheapest, yet will give high-quality performance. They are made with a good water valve that does not use a rubber washer. Some manufacturers use a tough ceramic disk; others use a hard plastic mechanism. Both are usually guaranteed for five to ten years.

If the valve seat does go bad, you merely replace it. That's a key thing to remember when you buy: purchase only faucets that come with replaceable valve seats. These better faucets and trim are also made with bodies of solid brass. Visit a plumbing supply showroom and you'll quickly see the difference between tough, well-made brass faucets and the cheap, light-weight kind. Though made of brass, the good guys will have a permanent copper-nickel-chrome finish. A tough protective coating is applied in the factory over the brass body. This gives a durable finish that doesn't tarnish and doesn't require frequent cleaning.

Their looks still may vary. They offer a wide choice of colors and degree of luster for virtually every decor, for example, a brushed satin finish, polished gold or chromium, or highly polished brass.

Good faucets for bathrooms include single-lever models, similar to those used in kitchens. Almost every brand is made with a replaceable valve. Some are made with a really top-notch valve, which means a few dollars higher cost, naturally. One of the best is the Moen single-lever. The main things to check when you buy a single-lever faucet are the length of the guarantee and how much repair work and expense are involved if the valve must be replaced. The style and design of single-lever faucets also vary but this, of course, is a personal thing.

Stepping up to top-of-the-line, deluxe faucets is not worth the extra money just for extra quality. These high-priced faucets are basically made with the same interior guts, basic valves, and construction as the middle-line faucets noted above. The fancy price goes chiefly for special styling and appearance, including models with a gold-plated finish. That can cost you as much as $300 more.

Shower Nozzles

A good-quality shower nozzle does not easily clog or corrode and has a flexible ball joint to adjust the spray direction. It should also have a volume spray control enabling you to obtain a fine or coarse spray. If you have hard water, a self-cleaning head is another good feature. The cheapest kinds have a rigid head, which cannot be adjusted, and little or no volume control.

Water, like energy, is becoming increasingly scarce and expensive so a shower nozzle that doesn't waste water may be important. Nozzles can spray as little as three gallons a minute of water, up to two to three times that much discharge. That obviously can make a difference in your water bill. So check water ratings, too. And definitely specify an adjustable nozzle that will allow you to vary the spray, as well as the water rate. An adjustable nozzle with a rating of three to seven gallons a minute is usually satisfactory.

12.
HOW TO CUT YOUR HOME ENERGY BILLS

There is probably no other aspect of building or buying a house where an ounce of prevention can pay off so well as when providing hard-core energy-saving measures when the house is built. With one exception, here are the main requirements for saving the most energy in your house. The exception is having the house trap free solar heat in winter, as described in the next chapter.

To cut energy costs, you simply chop away at the main energy drains in houses. According to the U.S. Department of Energy, they are:

1. Winter heating: 60 percent, approximately, of the average home owner's annual energy bill.
2. Spigot water heater: 18% (more with an electric water heater)
3. Cooking: 6%
4. Refrigerator-freezer: 3%
5. Air-conditioning: 3% (more with central air-conditioning)
6. Lighting: 2%
7. Clothes dryer: 1.5%
8. Other energy: 2.5%

Buying a factory house will give you a head start in the energy-saving department, compared with a stickbuilt house. Though all

Deep roof-overhangs shade the window glass from hot sun in summer, but six months later let sun shine in with welcome heat in winter, which is called having your cake and eating it too. New England Components/Techbuilt.

new houses must conform with the energy codes in 47 of the 50 states, as this is written, some codes are tougher than others. A factory house, however, is generally made to conform with the toughest energy code in a manufacturer's sales area. So no matter where you may live, a factory house is not likely to fall down on any energy-conservation measures, and is often leagues ahead of many codes. Depending on where you live, you could be taking potluck construction with a stickbuilt house. Because of its factory-built construction, a factory house is also, on the whole, better insulated and more tightly built to prevent air leaks than its site-built cousin.

You would know this instantly if you've ever worked outside on a bitter cold day. As skilled as a carpenter may be, he finds it difficult not to hurry a job on a frigid day in winter. Joints at the corners may show daylight; window framing, nailed up with a cold hammer held in an icy hand, could easily be more loose than it should. Hastily putting insulation in, he may leave gaps. Other mistakes occur at other times, such as during those dog days in summer when workmen are gasping for air and easily slip up.

A house built under a good roof and out of the weather in a factory is another story, as I've mentioned before. Tight joints are par for the assembly line. And workmen, under weather-protected conditions, can take the little extra time necessary to see that things like insulation are snugly installed. The occasional times when not, the slipup is usually caught by the quality-control inspector before the house leaves the plant.

But even the insulation and other energy-conservation specifications of a factory house also should be checked by every buyer, especially when the heating system and insulation are installed at the site. No matter what kind of factory house you buy, these are the important things to check.

Is There Enough Insulation?

Good insulation in the walls, floor, and under the roof of a house is the Babe Ruth of all energy-saving measures. It can by itself save fuel and cut winter heating bills by 30 to 35 percent, according to tests at the U.S. government's National Bureau of Standards.

Federal insulation standards apply to mobile homes and state insulation codes apply to many factory houses with the exception of those shipped with no insulation in part or all of the house. This leaves the amount of insulation to be installed open to a decision by the local builder or by you.

Insulation, by the way, neither stops the loss of house heat in winter, nor stops heat coming into a house in summer. Nothing stops it. But insulation is the best solution to date for *slowing* the escape of heat from a sprint to a crawl; the more insulation the more slowly the heat leaks out.

Insulation is measured by its R value. This stands for "Resistance to heat flow." A high R is good, a low one poor. The thicker the insulation, the higher the R value; e.g., 3 inches of fiberglass insulation has an R value of about 12 or so, depending on the brand;

Thick insulation is wrapped, like a blanket, around a manufactured house section. Cardinal Industries.

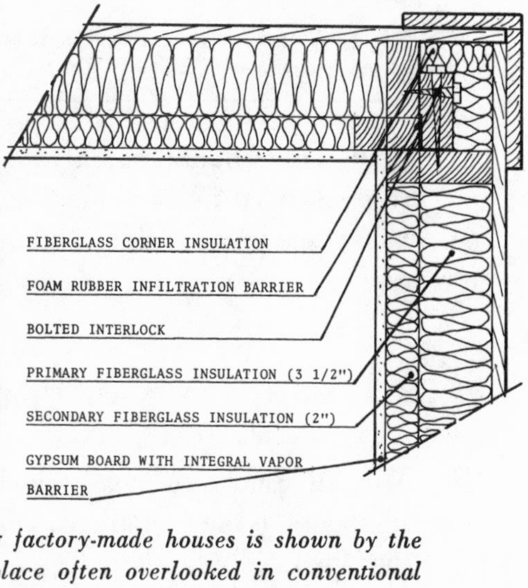

FIBERGLASS CORNER INSULATION

FOAM RUBBER INFILTRATION BARRIER

BOLTED INTERLOCK

PRIMARY FIBERGLASS INSULATION (3 1/2")

SECONDARY FIBERGLASS INSULATION (2")

GYPSUM BOARD WITH INTEGRAL VAPOR BARRIER

Special energy-saving treatment given many factory-made houses is shown by the copious corner-joint insulation. This is a place often overlooked in conventional houses and, as a result a common cause of heat loss due to air leaks. UGI's Capital Housing.

Actual construction of insulated corner shown in diagram.

6 inches an R of 20, more or less. The table that follows shows how much insulation to use according to the severity of the local climate measured by degree days.

The greater the number of degree days, the more cold weather in winter, thus the more insulation needed to keep heat from leaking out of a house. Annual degree days range from some 2,000 in the South for cities like Vicksburg, Mississippi and Shreveport, Louisiana; about 5,000 in the colder North for cities like New York City, Pittsburgh and Kansas City; up to 9,000 or more in the coldest northern parts of such states as Maine and Minnesota. The degree days for your area can be obtained by a call to the nearest weather bureau station.

INSULATION R VALUES RECOMMENDED FOR A NEW HOUSE *

Number of Degree Days	*Ceilings*	*Walls*	*Floors Over Unheated Space*	*Masonry Wall Insulation*
0-3,000	R-19	R-12	R-11	R-3
3,000-5,000	R-30	R-14	R-19	R-11
5,000-6,650	R-30	R-18	R-19	R-11
6,650 or more	R-38	R-18	R-19	R-11

* Adapted largely from "Thermal Performance Guidelines," published by the National Association of Home Builders.

The insulation put in your house should meet at least the amount recommended in the table above. That's whether you buy a factory house that comes with insulation installed in the factory, or after delivery at your site.

What Kind of Insulation?

Mineral wool insulation, which includes rockwool, glass wool, and fiberglass, is the best all-around insulation. It's a nonorganic material that's naturally bug-proof and fire-resistant. It's usually best to specify it rather than cellulose (a chopped-up newsprint product), which requires chemical treatment for fire-resistance and bug-proofing. Even when supposedly chemically treated, some cellulose brands are still not as good as they should be. Stick with mineral wool with a few exceptions.

There are styrofoam and urethane insulations, the two most efficient kinds commonly used in houses. They're more costly than mineral wool but can pay for themselves when used for board insulation, such as wall sheathing or foundation board insulation.

There is also urea-formaldehyde (uf) foam insulation which, alas, is subject to shrinkage after installation and sometimes causes a formaldehyde odor problem. It is very temperamental, especially when not properly put in a house. It's best to avoid the use of uf foam until some day in the future it has truly proved itself.

Insulating Glass

That means double-pane and triple-pane window glass. It's made of two or three parallel sheets of glass with a space in between each pair of glass sheets. Double-pane glass cuts the heat leakage, or loss, through windows by roughly 50 percent, compared with single-pane glass; triple-pane glass cuts the heat loss down to one-third that of single-pane glass.

In winter, the inside glass surface of multiglass windows is also less cold than the inside surface of single-pane. Among other things, fewer cold downdrafts are created on the inside of the insulating windows. In other words, insulating glass not only saves much heat, but it also means a more comfortable house interior.

Double-glass windows are recommended for all houses in a climate with 3,000 degree days or more in winter; with electric heat it's recommended down to 2,000 degree days, according to studies by the National Bureau of Standards. Triple-glass windows will pay for themselves in fuel savings and comfort in a climate with 5,500 degree days or more.

Insulating glass, however, is generally not worthwhile in the South for a house with airconditioning only. That means not worthwhile in air-conditioned houses in a southern climate with fewer than 3,000 winter degree days in winter, or, in a house with electric heat, fewer than 2,000 degree days.

Weatherstripping

This is a material made of fiber, plastic, or metal strips, installed around windows and doors to prevent air leaks in and out of a house. Such leaks are epidemic and are stopped only when weatherstripping is used to give a truly tight fit between a window and door and the stationary frame around each. Look at your present windows and you'll see how weatherstripping is used to prevent air leaks; or without weatherstripping, you'll see cracks and crevices of open light through which house heat can leak out in winter. Every window and door for a house obviously should come with good weatherstripping. Check this for the factory house you buy.

Choose the Lowest-Cost Heating Fuel

Use gas heat when possible. Depending on where you live, electric heat costs from two to *four times* as much as heating with natural gas. So who in his right mind will heat with electricity? Unfortunately, people buy houses without knowing that and without bothering to check on the relative cost of different fuels before they build or buy —and then pay the piper dearly.

Also depending on where you live, oil heat can run from 50 percent to more than 100 percent more than heating with natural gas. It depends on where you live because gas prices, like electric rates, vary from one area to another. Fuel oil prices are relatively uniform throughout the country.

Gas had not been available for heating many houses in the mid-1970s, but it has become far more so by 1979. Gas prices may also rise, but they have a long way up to go before gas heating costs as much as oil or electricity. Clearly, it's usually best to get gas heat

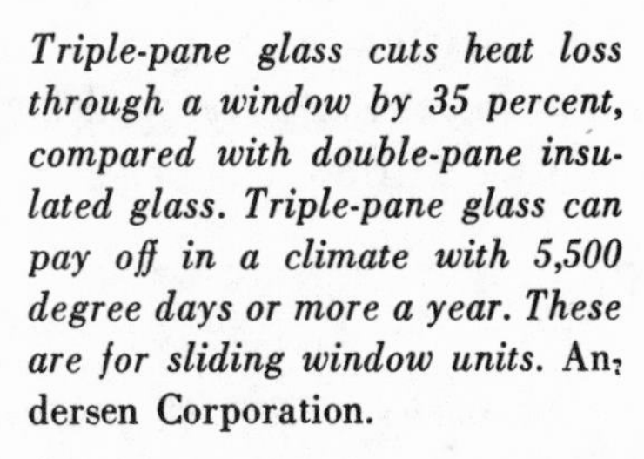

Triple-pane glass cuts heat loss through a window by 35 percent, compared with double-pane insulated glass. Triple-pane glass can pay off in a climate with 5,500 degree days or more a year. These are for sliding window units. Andersen Corporation.

A handsome fireplace is champagne for the soul in more than one way but, alas, don't count on extra heat from it—it can let much house heat leak up the chimney.

for lowest heating bills, and oil next. There are also minor options—coal and wood—which are discussed in a moment.

Comparing Fuel Costs

Here's how to determine the lowest-price heating energy in your area. It depends on the comparative price of gas, oil, and electricity. The following table, developed by the Small Homes Council of the University of Illinois, is based on the unit price of each type of energy.

HOW FUELS COMPARE IN COST

Electricity (¢/kilowatt-hour)	*#2 Fuel Oil (¢/gal.)*	*Natural Gas (¢/therm)*	*LP (propane) Gas * (¢/gal.)*
0.5¢	14.3¢	10.2¢	9.9¢
1.0	28.7	20.5	18.9
1.5	43.0	30.7	28.4
2.0	57.4	41.0	37.7
2.5	71.7	51.2	47.3
3.0	86.1	61.5	56.6
4.0	114.8	82.0	75.4
5.0	143.5	102.5	94.3
6.0	172.2	123.0	113.1
8.5	215.2	153.7	141.4

* For LP butane gas, use the price of natural gas per therm from the table. Source: The Small Homes Council, University of Illinois.

To use the table, determine the price charged for each fuel in your area. The price of gas, electricity, and oil for home heating should be obtainable by telephone calls to the local gas and electric companies and an oil dealer; LP (liquid petroleum) gas from an LP dealer. Ask for the local cost of gas and electricity for home heating and emphasize that you want the *actual* average unit price for home heating. Armed with the local heating cost of each energy, you now go to the table. It gives the equivalent prices for each heat horizontally across each line. For example, electricity at a cost of 3¢ a kilowatt-hour (kwh) for heating is equivalent to heating with oil at 86.1¢ a gallon, natural gas at 61.5¢ a therm.

If the cost of oil is less than 86.1¢ a gallon and natural gas less than 61.5¢ a therm, heating with either will cost less than electricity at 3¢ per kwh. If oil costs you, say 80¢ a gallon, it will cost less than heating with 3¢ electricity. If natural gas costs you 40¢ a therm locally, it therefore costs roughly one-third the bill for heating with 3¢ electricity and 86¢ oil.

Which energy is cheapest of all for heating your house? The table tells you.

A detailed description of fuel costs and other facts about heating energy is given in Circular No.G3.5, "Fuels & Burners," available for 25¢ from the Small Homes Council, 1 East Saint Mary's Road, Champaign, Ill. 61820.

Other Fuel Facts

A few other considerations should be taken into account when choosing among gas, oil, and electricity. Electric heat is the cheapest and easiest to install in a house. That's why it is put in so many new houses even though it costs more to operate. Gas is the next lowest in installation cost and oil the most expensive, especially because it requires an oil tank, too.

The actual operating bills with electric heat can be less than estimated because it's easy to install a separate thermostat in each room. As a result, the heat in any room can be easily turned off to save money, whereas this is far more difficult to do with gas or oil heat. In addition, an electric heating system usually requires less service and maintenance than oil or gas heat.

Moral: If the local cost of electricity does not make it *much* more expensive than gas or oil heat, electric heat can still make sense. In other words, if you estimate that electric heat will cost up to about 20 percent more than gas or oil, according to the table above, its other economies can offset its apparent higher operating cost. But be fairly sure that electric heat meets this guideline. For once it's installed in a house, it's difficult and expensive to switch to another heat.

There's less at stake choosing between oil and gas fuel. If you later find out that you chose a loser and the other fuel will cost less, it's usually easy to switch from oil to gas heat, or vice versa, with a minimum of expense. Only the burner mechanism usually has to be replaced. The rest of the heating unit remains the same. But do check that the furnace or boiler you get with your house can operate with either a gas or oil burner.

Coal and Wood Heating

Can you save with one of these? A lot of people are giving each serious attention. Together, they represent only about one percent of the home fuel market, but this may be changed. For the first time in years, coal sales by retail dealers to homes, churches, small commercial establishments, and so on, between January 1977 and January 1978 rose by one million tons. An indication of the demand for wood comes from Long Island, New York. In 1973, firewood sold there for about $45 a cord. Five years later, in December 1978, the increased demand had pushed the price up so much that people were lucky to find a cord for under $125.

Where coal is available for home use—and there is a serious supply gap in many metropolitan areas—it is likely to be less expensive than other fuels, according to the National Coal Association. An association spokesman said greater numbers of people probably would not be attracted to coal heating. You usually must buy a coal furnace, and once in use continually cleaning it out and discarding the ashes is no pleasant chore. He acknowledges that a small but growing number of people are buying and using small wood/coal stoves in living spaces to supplement, and in some cases replace, conventional central heat.

If you can pick up wood free, it's a distinct advantage. But if you must buy it, the cost could be as high as for gas or oil fuel. Taking into account the amount of wood required for house heat and the energy output of wood, Professor Jay W. Sheldon gives these comparisons in his book *The Woodburners Encyclopedia.**

Dr. Sheldon says that one cord of firewood at $100 a cord is equivalent to:

Electric Heat at 3¢/kwh	#2 Fuel Oil at 80¢/gal	Natural Gas at 60¢/therm	LP Gas at 55¢/gal

Dr. Sheldon's comparisons use a wood stove efficiency of 50 percent. Based on his figures, if your electric rate is 4¢ per kwh, wood at $100 a cord may turn out to be cheaper. But natural gas at, say, 35¢ a therm or oil at 45¢ a gallon would be nearly half the cost of $100 wood.

Five Other Ways to Keep Down Heating Bills

Get the proper-size heating unit. Heating units are measured by their Btu output (e.g., 86,000 Btu per hr.), and most important, bigger is not better. A Btu stands for British thermal unit, the standard measure of heat quantity. One Btu is equal to the amount of heat required to raise the temperature of a pound of water by one degree F. A heating unit—warm air furnace or hot water heating boiler—with an output greater than your house requirements will burn more fuel than necessary, because of its inefficiency.

Without being a heating expert, you can still insist that the heating system you get is not oversized, and also hold the seller responsible if it is. This cautionary note is especially valid now that manufactured houses coming off the line are very well insulated and tightly built. The result: these new houses require fewer Btu's of heat, in other words, a smaller heating system than the same houses built in the years before the United States energy crisis erupted.

Get a vent damper. With gas or oil heat, it can cut your heating bill as much as 20 percent. A vent damper is installed in the flue pipe that carries the exhaust gases from the heating unit to the

* Vermont Crossroads Press paperback, $6.95.

chimney. It closes when the heating unit goes off and thereby prevents residual heat within the system from escaping up the chimney, and also prevents cold air from coming down the chimney and chilling the heating unit.

A fail-safe device is included with an electronically operated damper. When the house thermostat calls for heat it first signals the damper to open. Only then does the burner fire start. If the damper malfunctions and refuses to open, the burner will not light. This prevents the possibility of flue gases dissipating into the house. In a competitive drive to provide the most efficient heating system, most heating unit makers now offer vent dampers with their equipment. You can save money by specifying one with your heating unit. Electric furnaces have no chimney vent so have no need for a vent damper.

Install an electronic spark ignition. This eliminates a continuously burning pilot on gas furnaces. A thermostat-controlled spark ignites the pilot, which then lights the burner. When heat is no longer needed, the gas to the burner and pilot is cut off. Most new gas heating units have this device installed as a matter of course. It may cut your fuel bills no more than 1 to 4 percent, but every little bit helps.

Specify an automatic-clock heating thermostat. It can cut fuel bills 7 to 15 percent, depending on your climate. It automatically tells the heating system to doze off and provide less heat to the house at night when you sleep. It's also set to turn up the house heat at whatever time you want more heat when you awake. It will pay for itself in savings in a year or two. One kind provides a double setback: one at night; the other for daytime when all members of a family are regularly away from home. Each degree of setback in winter, day or night, will save approximately 3 percent of the energy used at the higher setting. The same 3 percent saving applies to raising the night setting in summer with a cooling system.

Be sure all vulnerable heating ducts and pipes are insulated. This means any heating or cooling duct or pipe that travels through an unheated or non-air-conditioned part of your house, such as a cold crawl space under the house, or a hot attic above. Leaving such conduits with no insulation means heat losses in winter to the unheated space; and, in summer, cooling losses to the same space; thus increased energy bills. It's wise, in fact, to insulate *all* cooling ducts in an air-conditioning system, if only to prevent condensation on the outside of the duct. Heat ducts or pipes running through a conditioned space do not benefit much from insulation.

Get a high-efficiency water heater. Because the water heater that provides hot water to the kitchen, bathroom, and laundry is the second biggest guzzler of energy, topped only by the heating unit, getting a good one can obviously save more than a few bucks on monthly energy bills. Here are pointers for getting a good self-contained water heater and storage tank, the most common kind. It operates independently of the house heating unit, though both may vent into the same chimney.

Specify a water heater operated by natural gas, unless gas is unavailable. Like central heating, a gas water heater is lower in cost to buy and operate than an oil or electric water heater. Even in a house with electric central heat, a gas water heater can be more economical than an electric one. Exceptions are in low-cost electric power areas, such as the Tennessee Valley Authority (TVA) and the Pacific Northwest.

Get a high-efficiency water heater, which means a superinsulated tank. It costs a little more to buy than standard models, but its high efficiency will repay that extra price many times over in annual operating savings.

Have a vent damper installed with the water heater (except for an electric water heater which is ventless). In fact, if you have a furnace with a vent damper and a water heater without one, the effectiveness of the furnace damper will be reduced, assuming both units went up the same chimney flue. As with central heat, a vent damper on a water heater can cut monthly energy bills as much as 25 percent.

On the other hand, "Don't buy a water heater with an electronic spark ignition," says a spokesman for the Hydronics Institute, trade association for the hot water house heating industry. He says, "A burning pilot light maintains water heat for hours, much reducing the number of times the burner is fired and is worth the small amount of energy it expends."

Appliances

Cook with gas. Cooking with an electric range can be from two to four times more expensive, again depending on the relative cost of each where you live.

A gas range and oven. It should have an instant ignition device. Like electronic furnace ignition, this means no constantly burning pilot, and it can cut gas cooking costs as much as 10 percent. Other energy savers are thermostatically controlled top burners which maintain pot temperature with only the heat needed. Admittedly, a self-cleaning oven is not an energy saver, but it is a superior work saver, and you ought to give yourself a break once in a while.

A microwave oven, while not for all-around cooking, saves energy for what it cooks. In baking four potatoes, for example, it uses 61 percent less energy than a standard electric oven requires.

The refrigerator. High-efficiency refrigerators with improved insulation are practically standard now. They save 10 percent over older models. Manual defrost saves more energy than a refrigerator with automatic or fully automatic defrost, but there's a catch. If you fail to defrost the manual model, frost buildup will force the motor to run harder and more often, which could drop its energy efficiency below an automatic unit. And get a refrigerator with a "power saver" switch. When the house humidity is at a low level, you can turn off the switch and save energy. It's also good to have on a separate freezer.

Dishwasher. Be sure to get a dishwasher with a control that eliminates the drying cycle. When you turn off the cycle, the dishes air dry. The control gives you the option of fast drying when you need the dishes in a hurry. Also get a built-in heat booster. This raises the temperature of incoming water to germ-killing levels, while letting you maintain a money-saving lower-temperature setting on the hot water heater.

Clothes dryer. Get one with a "drying sensor." This gauge tells when the clothes are dry. If all moisture is gone before the drying cycle ends, the sensor turns off the machine.

Clothes washer. Researchers have found that 80° F water will clean clothes as well as 120° F water—the difference between "warm" and "hot" settings on machines. Until 1977 most manufacturers set a mixture from house lines for "warm" at 50 percent hot and 50 percent cold. Testing revealed that this 50/50 mix produced water warmer than 80° F. As a result, some makers now offer washers that make a mix of around 40 percent hot and 60 percent cold. The water temperature falls far closer to 80° F, and hot water needs for the clothes washer are cut by 20 percent.

Central Airconditioning

Specify a high-efficiency airconditioner. That means one with an "Energy Efficiency Ratio," or EER, of 7.5 or better, and preferably 10. The higher the ratio, the more cooling the unit provides compared with a standard airconditioner drawing the same amount of electricity; hence lower operating bills. Even though a high EER conditioner cost more initially, it will pay for itself in savings in a few years; the savings that continue every year afterward is money in the bank.

Be sure that the cooling unit is sized properly for your house. Like the house heating unit, an oversized cooling unit is an inefficient energy waster. In addition, too big a cooler fails to perform an important summer job: dehumidifying the air. This advice also applies to room units. If possible, have the condenser (the outside portion of a split system) installed out of the sun's direct rays.

The Heat Pump

If you live in an area with low electric rates, or electricity is the only practical power source and you want heating and cooling for your house, a heat pump is a good energy-saving choice. It operates the same as central cooling in summer, pumping hot indoor air out and leaving cooler air behind. In winter the machine reverses itself and extracts heat from outdoors to deliver it indoors. Produced commercially for over 30 years, the heat pump's performance today is excellent throughout the South, good in the temperate North but still questionable in a climate colder than Washington D.C. (4,333 heating degree days), according to the National Bureau of Standards engineers. Some manufacturers claim that their heat pumps will perform well in colder climates. But before swallowing such a claim, I recommend that you check on the actual performance and the winter bills for the same brand units in your area. Talk to people who have them.

The big advantage of a heat pump is that it heats with electricity in winter at significantly less electrical cost than regular electric heat. Its cooling cost in summer is the same as cooling with a conventional electric airconditioner. If you choose a heat pump, judge it like a central airconditioner. Get one with an Energy Efficiency Ratio of at least 7.5 or higher; the higher it is, the lower your operating bills in summer. And it should have a Coefficient of Performance (COP) of about 2.0 at 15° F to 20° F outdoor temperature in winter, to 2.8 at 45° F to 50° F. Those two ratings will mean comparatively low operating bills in winter.

Cutting Lighting Costs

Watts are the guide to power use. A 100-watt incandescent bulb draws two-and-a-half times more electricity than a 40-watt incandescent bulb. This fundamental fact makes fluorescent tubes and high-intensity-discharge lamps (HID) very attractive because they provide more illumination with less power, watt for watt, than incandescent bulbs.

For example, a 75-watt incandescent bulb throws out 1,180 lumens (a measure of light output). A 30-watt deluxe fluorescent tube delivers 1,530 lumens, or 30 percent more lighting power from less than half the energy used. Similarly, a 100-watt incandescent produces 1,750 lumens, while a 100-watt deluxe white mercury (HID) lamp glows with 4,000 lumens.

On a one-to-one basis, fluorescents and HID sources cost more than incandescents, but they last much longer. The 75-watt incandescent bulb has an average life of a mere 750 hours; the 30-watt deluxe fluorescent, 15,000 hours, and the HID lamp, 18,000 hours. Overall, fluorescents are clearly much less expensive to buy and use.

Many people cannot imagine using fluorescents or HID lamps in spaces other than a workshop, laundry, kitchen, and bath. The fact is, decorators often select these long-life sources for living areas, though they are careful about selection. They choose deluxe units only. The glow, for example, of a deluxe warm white fluorescent is quite similar to that thrown off by an incandescent bulb.

Where you choose incandescent bulbs, have them installed with dimmers. Dimming switches not only give you means to save energy, but a way to produce the glamour of changing moods in each living space.

What About the Fireplace?

Unfortunately, the preenergy-crisis fireplace wasted more house heat up the chimney than it gained by radiating its heat to the house, though it may not have seemed that way. That wonderful draft up the flue not only drew out the smoke, but also let a lot of indoor furnace heat get away up the chimney. When not used, and even with its damper closed, a fireplace can let house heat continue to escape up the flue past cracks between damper and firebrick surround. Such losses can be minimized with a fireplace for a new factory house if you get one with:

1. A damper with a positive close.
2. A glass fire screen (alas) to prevent loss of furnace-delivered heat. It gives total assurance against the danger of fire from flying embers, but still provides the radiant heat given off by the fire.
3. Air-circulating vents. At a little extra cost, these are fireplace forms with air-flow ducts that capture more of the fire's heat. They deliver it safely to the room served by the fireplace, and, in some cases, to adjoining rooms.

13. HOW TO USE FREE SOLAR HEAT AND CUT HOME ENERGY BILLS

A couple I'll call Jackson made their bed right when they bought their factory house not long ago. They made sure that the house would face south on their land and trap big loads of solar heat in winter. As a result, their fuel bills run sharply * lower than those of friends who live in a similar house across the street. In summer some of the same measures that cut their heating bills in winter also help make their house easier to cool and hence lower air-conditioning bills.

Big fuel savings have been documented in manufactured houses that are specially designed for what's called "passive" solar heat. That awkward phrase, apparently coined by a bureaucrat, means opening up a house properly to take full advantage of all of that solar heat up there that's free for the asking without an expensive mechanical solar heating system. In other words, without a roof

* That's still another percentage tossed at you in this book and it may seem that so many different savings mentioned should mean that so much can be saved, if you're a good boy or girl, that by now your energy bills will be down to zero. I only wish so. The catch is that each time energy bills are cut, there's less of the cake left for making future cuts. Suppose, for example, you cut energy bills with Plan A by, say, 20 percent. Next there's Plan B that can save 15 percent. But that's only 15 percent of the remaining 80 percent left after the first savings made with Plan A. It cannot be 15 percent of the total original bill when you started cutting it. I also assure you that the various percentages in this book are not pulled out of thin air; they are from good sources, such as the FHA and the National Bureau of Standards.

collector, pipes, valves, etc., which is called—brace yourself again—"active" solar heating.

Independent engineers did a study of one house, for example, made by Green Mountain Homes of Royalton, Vermont, designed for maximum passive solar heat efficiency. The study showed that during the bitter cold Vermont winter of 1977-78 (more than 9,000 degree days) the house was heated for a total eight-month energy bill of only $249. Free solar heat did so much of the heating which kept the need for costly energy down that low. During a typical Vermont winter (approximately 8,000 degree days), the study shows that the house is heated for an average of $175 total energy cost. That means more than 50 percent saved on heating fuel, which isn't bad.

In addition to facing a house south, two other requirements are necessary to trap a lot of free solar heat in a house in winter: plenty of window glass facing the south sun, and ways to circulate and store the solar heat inside.

About the only other way to achieve the same end is to build your house on a giant turntable that can be revolved at will, turning it around to drink in the southern sun in winter, its back to the cold north; and come summer, turning it around 180 degrees so its back is to the hot south and it faces the cool north.

Do that and there's little need to read the rest of this chapter. If you can't swing a turntable, here is how you can profit considerably by properly orienting your house on its land. The annual energy savings that result not only run as high as 60 percent, compared with locating the same house with a potluck exposure and design, as most houses get. Your house also will be brighter and more cheerful in more than one way; e.g., in higher future resale value.

The Ideal Exposure

The ideal exposure for most houses in the northern hemisphere is broadside to the south, because in winter the sun rises in the southeast and sets in the southwest, as shown in the diagram. In summer, the sun rises in the *northeast*, travels a much higher arc across the sky, being almost directly overhead at noon, and sets in the northwest. A few scientific facts emphasize what this means:

1. Only rooms with windows that face from southeast to southwest receive much warm sunshine and solar heat in the dead of winter. In case the importance of that escapes anyone, window glass can let an enormous amount of sun heat pour into a house. It's the greenhouse effect, the phenomenon that makes temperatures inside a greenhouse kitchen-hot.

2. Every room on the south side of a house receives five times as much sun heat in winter as in summer.

3. East and west rooms, on the other hand, receive six times as much sun heat in summer as in winter.

4. Rooms and windows facing north receive no sunshine at all in winter and only a little in summer; the farther south you are, the more sun is shed on the north side in summer.

Clearly, a house with a lot of window glass facing south can sop up much free solar heat in winter. Enough heat pours in that your furnace can take a rest for much of the day—hence major fuel

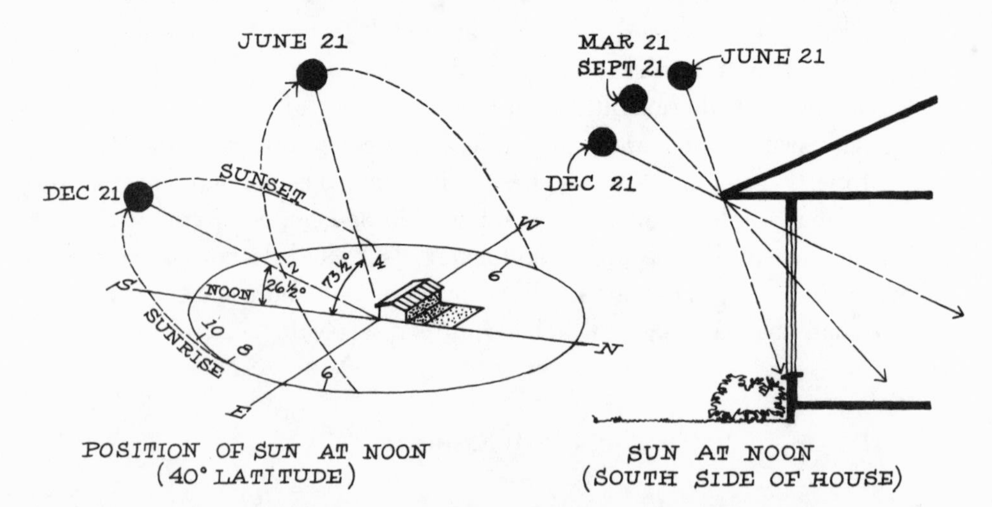

A house or its main windows clearly must face South or close to it to trap any real quantity of solar heat in winter. This diagram applies to all houses in the northern hemisphere; it's the reverse in the southern hemisphere. But in summer the same south-facing windows can be easily shaded from hot overhead sun with roof overhangs, since the sun comes in from a higher overhead angle. Shading can also be provided, of course, by trees or awnings. The Building Institute.

savings. That's done with no mechanical solar heating system; in other words, no elaborate solar collector panels on the roof and no expensive system of pipes and ducts and other complicated (and service-prone) accessories for capturing and distributing solar heat the mechanical way.

In 1979, as this is written, many experts have just about concluded that mechanical solar heating systems for houses are still far from practical. They are not working out any better, alas, than our long-term efforts to extract pure water from our vast oceans. Solar water heaters for houses were the only solar units with promise in houses, and these can pay off only as an alternative to high-cost electric water heaters. That means it pays off only if you cannot use a gas or oil water heater, both of which should be considerably cheaper to operate. In short, mechanical solar heating systems are not only still far from being proved in houses, but also, unfortunately, may not be for many years, barring a major technological breakthrough. Until that happens, no one can honestly recommend building or buying a new house with a mechanical solar heating system. That includes saying no to virtually any kind of mechanical solar heating optional system that is offered by a home manufacturer.

How to Use Free Solar Heat in Your House

This is a wholly different ball game and it's eminently practical—and cheap.

1. Try first for a site on which your house can be given a good southern exposure. The ideal lot is one on the south side of the street. Land on any side of the street often can work well if it offers flexibility. It will also help, of course, if you have a good view toward the south. Remember that it's perfectly all right for the back of a house to face the street. It helps, too, if the back is reasonably attractive.

2. Choose a house that can easily face south, the Mecca direction for solar heat. It doesn't have to be due south. Tests show that excellent solar heating can result when a house faces up to about 30 degrees east or west of due south. If necessary, have windows added on the long south side, while eliminating as many as possible from other sides of the

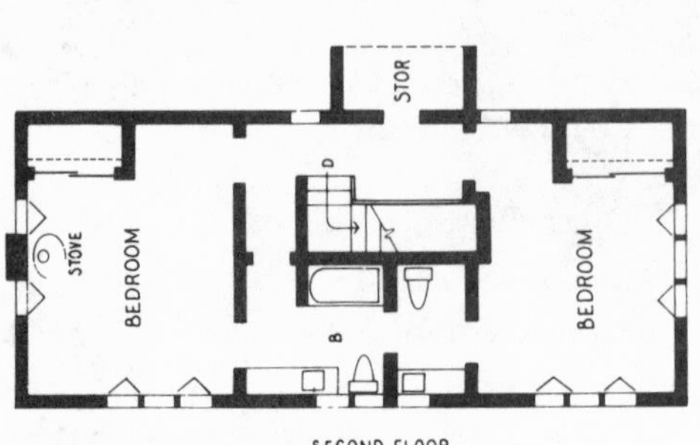

SECOND FLOOR

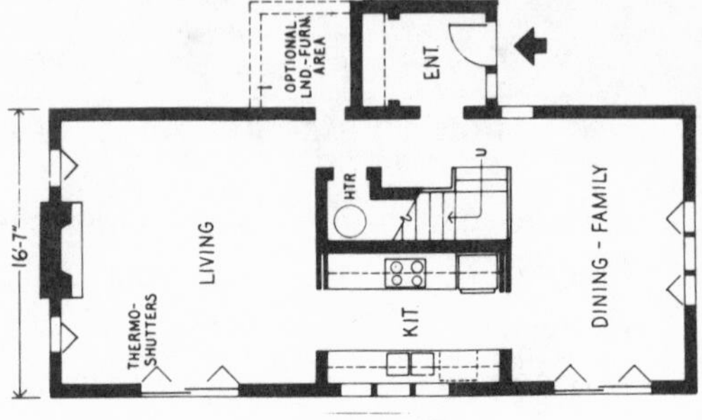

FIRST FLOOR

Winter heating bills are cut by more than 50 percent as a result of free solar heat, also called sunshine, pouring through the windows of this house. Located in cold Vermont, it is designed to trap and use solar heat without an expensive, mechanical solar heat system. Green Mountain Homes.

Any house can trap natural solar heat merely by use of the greenhouse phenomenon—expose plenty of window glass to the warm south sun in winter. Roof overhangs shade the same glass in summer; the further north the house, the deeper the overhangs required. Haida Hide, Inc.

house. Most home manufacturers will gladly make window changes in their houses.

Actually, the house itself can face any direction. The important thing to remember is to have the windows of your daytime living areas (kitchen, dining, and family rooms) facing south for sunshine, light, and sunshine to flood in. A south exposure is less important for the living room, unless you use it a good deal during the day; many people use it mostly at night.

Bedrooms obviously have less need for daytime sun and are best on the north or east. On the east, you may welcome the bright sun and free heat flooding in first thing in the morning. Bedrooms on the west can get furnace-hot in summer by the time you are ready for bed. Try to locate the garage or carport on the west as a sun shield in summer, or on the north as a wind shield in winter.

Little or no window glass should be on the vulnerable, wind-lashed north side of your house to keep winter heat losses from the house down to a minimum. That includes a wall exposure ranging from north-east around to west, depending on the prevailing winter wind where you live.

Good weather protection in winter can be had with a windbreak of evergreen trees on the north, as is frequently seen near farmhouses in the wind-scourged Great Plains. If you live in the South, a south orientation may be less desirable. You will have more shade if a house faces north, with the patio on the north or northeast.

3. Have good summer sun control built into the house. The beauty of facing windows to the sunny south is that south windows are comparatively easy to shade from hot summer sun. Hot sun blasting down on window glass in summer is, in fact, the second largest cause of high air-conditioning bills. (The first is hot sun on the roof all day long.) Shading windows from direct sun not only can mean sharply reduced air-conditioning bills in summer, but also a cooler house without air-conditioning.

Roof overhangs are one of the best ways to shade south-facing windows from direct sun in summer. A roof overhang can easily keep out the summer sun. But in winter, the sun sails under the overhangs and into the house through the same windows because the sun is shining in from a much lower angle in the sky, as shown in the illustration. The farther north a house is located, the deeper the overhang necessary to keep out the sun in summer. The manufacturer of the house you buy can tell you how deep south roof overhangs must be for good sun shading in your latitude.

4. Keep the sun heat inside after you've caught it. This calls, first of all, for a house that is very well-insulated and, except in the warm

South, has double- or triple-glass windows. Solar heat from the sun can pour in through multipane glass virtually as easily as through ordinary single-pane glass, so double or triple glass is fine for trapping solar heat. But heat inside a house has a different wave length. It can leak out quite easily through single-pane glass, the thinnest, and least energy-conserving kind of glass, but not so easily through multipane glass. Multipane glass windows generally do not pay in the warm South for either heating or central airconditioning, as noted in the previous chapters.

5. All other things being relatively equal, use forced warm air heat. That means, of course, central heating with a fan that circulates the warm air heat around the house. The same blower and ducts are used to spread solar heat around the house; that is from the warmed-up rooms on the south side to other rooms. Even when no heat from the house furnace is needed, the fan can be turned on to circulate solar heat in the south side of the house to the rest of the house.

6. Store excess solar heat coming in during the day for use at night though, unfortunately, this is easier said than done. Heat-absorbing materials inside the house are used to soak up sun heat pouring into the house. Later when the sun goes down, the same materials will release their heat to help keep the house warm. Such materials include brick, stone, or concrete floor on the sun side of the house, or drums of water (though these may not be the prettiest things to have in a house).

The catch is that, so far, such heat-storage ideas have not been sufficiently developed and tested for use in houses. A solid concrete floor is apparently working well in the houses with passive solar heat mentioned early in this chapter but, unfortunately, one sparrow doesn't make a summer. But by the time you read this, some heat-storage methods may be ready for sucessful use in houses.

To find out, ask house manufacturers, their salesmen, and other people about them when you shop for a house. What are good things to build into a house to trap and store solar heat? Have they proved themselves? How can you use them in your house?

Don't take no for an answer when you request a house that can make maximum natural use of solar heat. It's been around for a long time. Smart builders, architects, and others have been using it in houses throughout the world for hundreds, if not thousands, of years. No reason why the rest of us can't do it today.

14.
OTHER IMPORTANT FACTS ABOUT FACTORY HOUSES

Here are a few final tips and a review of key facts about manufactured houses that may have been omitted or inadequately presented earlier in this book but deserve special mention:

Choose the Right Land, or Site

This is the key, obviously, for making it easy for a house to face south and be easy and cheap to heat and cool, particularly if you buy a factory house in a development or a mobile home in a mobile home park. The worldwide energy crisis has caused a major reappraisal of house design in the United States. The orientation of a house in relation to the sun and cold winds in winter is now of great importance, assuming you don't want to go broke paying high energy bills in the future.

Remember that a house doesn't have to face due south but can face up to 30 degrees or so east or west of due south. Best of all is a lot on the south side of the street with the house and its big windows facing the rear of the lot.

Should you buy a private lot of your own, its geographical location obviously can make a big difference in your living and in the future resale value of your house. In addition to the high priority put on a southern exposure, here are things to check on these scores: The local zoning should assure the area of maintaining its residential character. No one wants a gas station or fertilizer plant to go up across the street. There also should be good local transportation, good schools if you have children, convenient nearby shopping, churches, parks or other recreational facilities, and no nuisance neighbors, such as an airport, noisy highway, or disco bar next door.

The cost of improving the land can be much reduced if the utilities (gas, water, electricity, and sewage disposal) are readily available. Can each be had at minimum expense? What about water runoff and drainage from the land?

Land improvement cost also can be reduced if your house can be put down near the front of the lot. Then the utility lines from the street to the house (which you pay for) are kept short. You will also save on the cost of a driveway and walks.

Use a Good Local Builder

Even the best factory house package can be easily sabotaged if assembled at your site by a hack builder (who can seem sweet and dependable when you first meet him). What's more, if the house is botched, the manufacturer is justified in claiming no responsibility.

One couple, for instance, who bought a factory house hired a local builder recommended by a relative to erect it. He had offered to do the job at a lower price than two builders from the manufacturer's list of recommended builders.

Completion of the house was botched and the couple complained to the manufacturer. The manufacturer pointed out examples of sheer carelessness, including mistakes that showed that the builder hadn't even bothered to read the assembly instructions. "Obviously, we are not responsible," the factory representative told them. "Read your contract," he said.

The couple could not get the guilty builder to make amends; in fact, he couldn't be found. Moral: Read your contract carefully and also be sure that the local builder who puts your house package together is really good, even if you must pay a good guy a higher price than a low-bid builder with unproved ability. That extra price is, in effect, an insurance premium paid for your protection.

Warranty Insurance on Your House

About the only way to protect yourself against serious defects developing in any new house today is to buy from a builder who participates in the Home Owners Warranty (HOW) program sponsored by the National Association of Home Builders. That includes builders who put up and sell factory-made houses.

Not all of them may offer this insurance because it's relatively new. But it's well worth asking about because it gives you an excellent 10-year insurance plan at relatively low cost. The coverage varies from virtually full protection on any faulty part of the house for the first year; protection against serious defects in wiring, plumbing and structural elements for two years, and a warranty against major structural defects for ten years. The cost is $2 per $1,000 of sales price; thus $130 for a $65,000 house. Some builders absorb this cost in their overhead; in other cases the buyer pays it. The warranty is transferred to the new owner when a house is sold.

Perhaps the main advantage of a HOW warranty is that builders who offer it try harder. It encourages them to impose greater quality control on their site construction and, as a result, less chance for oversights and possible defects showing up in a house later.

Building Your Own House

Don't bite off more than you can chew. Building a house from a factory package, or kit, can put you leagues ahead of the game, compared with building a stickbuilt house starting from scratch, as mentioned earlier. But it's still not child's play, even for a skilled do-it-yourself person. Putting up a whole house is the do-it-yourself equivalent of climbing the Matterhorn, the supreme peak.

Figure that it may take as long as a year, if not longer, depending on how much time each week you can give to it. Unless you're a triple-threat construction man, some of the work (excavation and foundation, heating, plumbing, and other such tough chores) probably should be subcontracted to professionals. In short, be realistic about your capability. Many home manufacturers, in fact, discourage sales to do-it-yourself buyers. For one thing, they don't have the time or manpower to nursemaid a buyer through the complete birth of his house, and for another they want to avoid trouble and mistakes made by such buyers. Certain makers do, on the other hand, specialize in sales to do-it-yourself buyers.

Two good books on the subject are: *How to Build Your Own Home,* by Robert C. Reschke, Structures Publishing Co., $7.95 in paperback, $14 in hardcover, and *So You Want to Build a House,* by Peter Hotton, Little-Brown & Company, $6.95 in paperback. Both can be had by mail from The Building Institute, Dep. FH, Piermont, N.Y. 10968. Add $1 mailing cost for each.

Getting a High-Quality Factory House

In chapter 8, mobile homes, I mentioned why a mobile home and all other houses can pass tough building codes but still lack uniformly high-quality construction. It's because a building code applies only to those parts of a house that affect the health and safety of people. The quality of much of the rest of a house can be borderline, if not low, and your health and safety will not be in jeopardy.

The water heater, for example, may be absolutely safe in operation but not very energy-efficient, thus sky-high energy bills with it. You may have to work overtime to keep kitchen and bathroom floors looking good because of low-quality flooring. Other parts of the house quickly get frayed and unattractive, but that's par for the course.

A truly efficient water heater, good flooring and other such products will mean low upkeep and long life, but they are not covered by building codes. They also tend to be the parts of a house that a home manufacturer, like a local home builder, is not likely to spend more money on than necessary for high quality. A good part of the reason is that home buyers don't know the difference and will not pay extra for a house of uniformly high-quality construction. Most buyers will go down the street, or to another home manufacturer, and buy a cheaper house. In self-defense, therefore, manufacturers say they are compelled to concentrate on using the lowest-cost parts and products to keep down their house price tags. Naturally, that's no formula for good quality, which costs more.

The extra cost of high-quality materials is often less than many people think. Paying extra for really good-quality products and materials where it can count, as noted below, can increase the cost of a typical house by a mere 5 to 8 percent, according to a study by *House & Home* magazine. That will mean a relatively small increase in your down payment and monthly mortgage payments, but this will be offset many times over in reduced upkeep and house operating costs. A good example is getting a permanent-finish outside wall surface. This can save the cost of repainting, or a periodic expenditure of up to several thousand dollars.

You be the judge. If you expect to stay in your house less than five years, paying extra for a better house may be a questionable investment, except that it can make the house sell for you at a higher price. If you expect to own it more than five years, the longer you're in the house, the more a high-quality house makes good financial sense. You'll get your money back compounded, not only in savings in upkeep and service, but also in pleasure.

Here are the most likely places to check for quality and, if necessary, to spend a little more money for better-than-ordinary quality products furnished with a house or mobile home you may buy. This doesn't mean, by the way, that you must shell out high prices for top-of-the-line, Cadillac quality. Many products are made in three or four steps up in quality, and merely stepping up to the second level above dead bottom can make a big difference—and at comparatively small extra cost.

Kitchen fixtures, appliances, cabinets, and flooring: The extra price for a high-efficiency refrigerator and a good dishwasher will pay for itself in monthly operating savings and reduced service, no question about it. Good fixtures, good cabinets, and good flooring will mean a welcome reduction in cleaning and upkeep, as well as prolonged attractiveness.

The bathroom: Get a good sink (lavatory), tub, and toilet, which means adequate size as well as solid design and construction, as described in detail in chapter 11.

Furnace and water heater: A high-efficiency furnace and water heater should be top priority requirements. Without them you're a sitting duck for high, higher than necessary, energy bills.

Exterior walls: An outside wall, like long-life brick, stone, or vinyl and aluminum siding, can save a small fortune over the years in repainting costs.

Windows and doors: Getting good quality here includes specifying double- or triple-pane window glass in the North, and tight-fitting windows and doors, i.e., with good weatherstripping all over.

Financing a Factory-Made House

As with buying a stickbuilt house, this means getting a mortgage loan with a down payment and monthly repayments that you can live with and no small-print hooks. The importance of shopping for one cannot be overemphasized. Yet some people will spend more time shopping for steak and potatoes than for the home loan document that can tie them up for tens of thousands of dollars for years and years.

Sometimes, of course, just finding a bank or other lender who will give you a mortgage can be difficult and, like a beggar, you must take what's given and that's that. But at other times, lenders have money for mortgages coming out of their ears, and mortgage shopping can be the happy equivalent of gunning down targets in a shooting gallery. You can take your pick. It depends on the state of the national market for mortgage money; sometimes there's plenty of money available for mortgages, other times not.

In any case, call banks, savings and loan associations, mortgage brokers and anyone else a manufacturer may recommend for a mortgage loan on his houses. There are the special things to find out from each lender: obviously the mortgage interest rate, down payment, and number of years to repay. In addition, ask about such small-print provisions as a prepayment penalty fee if you repay any portion of the mortgage ahead of when it's due. That includes paying it off if you sell the house before the mortgage has been repaid. It's best when there's no prepayment penalty after a year or two from the date you get the mortgage.

Does the mortgage lender have the right to raise the interest rate? He should not, with one exception: with a variable-interest mortgage on which the interest rate can go down as well as up according to national interest rates for all money. Do you have to pay "points" for a mortgage? That's a special charge that can amount to hundreds, if not thousands of dollars. And very important, what will the total closing-costs bill be? That's for the fees and special charges related to obtaining a mortgage; e.g., title insurance, mortgage "commitment" fee, and so on.

But also remember that often the terms of a mortgage loan and its closing costs are sometimes negotiable, particularly when a banker's ears are brimful of money. In other words, you can bargain for better terms than offered. Do you want a lower interest rate? Smaller down payment and cheaper closing costs? Ask for them. Don't be shy. After all, it's your money that's at stake. And merely speaking up for the best deal can earn respect, assuming you perform in a businesslike way and with no emotional hysterics.

PART 2
DIRECTORY OF HOME MANUFACTURERS

The names and addresses of more than 200 home manufacturers are listed here according to the kind of houses made: modular, panelized, precut (including log and dome houses), and mobile homes. The facts given here are based on information received from each at the time this book was written.

Remember, though, that a maker's houses are subject to changes. Models are continually being added, others dropped, and still others being revised. As a result, the accuracy of information given here cannot be guaranteed. What's sold today is not necessarily sold tomorrow.

Little or no information is given about some companies because of a sparse reply or no reply to my requests for information about their houses.

The size of the houses sold, in square feet (sq. ft.) of interior floor space, and a company's shipping distance are given where available, as these are two obviously important facts. As noted in chapter 3, the cost of a house is directly related to its size in square feet; the more space in square feet, the higher its cost in a straight line proportion. House costs, less land, range from about $16 per square foot for mobile homes; and for other houses, from about $20 to as much as $40 per square foot, sometimes more, for a finished house. Most makers can tell you the cost per square feet for their house packages, or kits, which of course doesn't include their on-site construction cost. Many are also glad to give you the estimated total cost of their houses completed by a professional contractor.

Prices are not given for all houses in this directory because they can be highly misleading for several reasons. Prices for manufacturers' packages obviously vary all over the map according to how much of the house is supplied by each manufacturer, and this can vary considerably, as noted variously in this book. Comparing house packages by their prices is as bad as comparing apples with figs and prunes.

Prices of a factory house package not only do not include such sitework costs as excavation, foundation and utilities, but these costs also can vary according to the size and type of factory house. Finally, house prices have been continually rising so what they are when this book is written is almost certain to be obsolete at the time it is published and read. Obtaining up-to-date factory house prices therefore requires asking manufacturers for them when you shop for a house.

The size of a house in square feet can also tell you whether it will fit your requirements. A house with 1,500 square feet of interior space is roughly the minimum required for a family of four or five. The larger the house you can afford over this minimum, obviously the more breathing space for each member of your family. Knowing about square feet of space in a house is like knowing about horsepower in a car you may buy.

Because most manufacturers will ship no more than a specific distance from their plants, their houses are not sold beyond that shipping radius. Some, however, have two or more plants so knowing the closest plant location can be helpful.

The larger, more established home manufacturers tend to be members of the National Association of Home Manufacturers (NAHM), 6521 Arlington Boulevard, Falls Church, Virginia 22042, the home manufacturers association for all but mobile homes. Dagger shows companies that are members of the NAHM (†). Its counterpart in Canada is the Canadian Home Manufacturers Association, 27 Goulburn Avenue, Ottawa, Ontario K1N 8C7. It can provide information about home manufacturers (other than mobile homes) in Canada. The trade associations for mobile homes in the

United States and Canada are cited later in the introduction to the mobile homes section.

Modular Houses (also called Sectional Houses)

A modular house is virtually a complete house when it leaves the factory. As described in chapter 4, it is shipped in one or more sections, each being 12 or 14 feet wide, one-story high with roof in place, and up to 60 feet long. In short, it's a three-dimensional package, with two or more sections connected at the site to form a finished house. After connecting the sections, usually all that's needed for you to move in is connecting it up to outside sources for energy, water, and sewer.

Active Homes Corporation, P.O. Box 127, 7938 S. Van Dype, Marlette, Michigan 48453. Houses up to 2,000 sq. ft.; also makes panelized houses and modular motel and apartment units; sells in Michigan only.

Arabi Homes, Inc., P.O. Box 117, Arabi, Georgia 31712.

Capital Housing, Inc., P.O. Box 326, Avis, Pennsylvania 17721.

Cary-Way Building Company, P.O. Box 26937, Houston, Texas 77207.

Cedar Chalet Div., Leisuretime Products, Box 445, White Pigeon, Michigan 49099. Also makes panelized and A-frame houses. Most of its models are one-story ranch houses up to approximately 1,625 square feet of floor area; sells through dealers located from the east coast west to Wyoming.

†***Century Homes,*** Route 2, Box 71, Maxton, North Carolina 28364, a division of R-Anell Homes, Inc. Houses range in size from 960 sq. ft. to 2,064 sq. ft.; sold in the South, from Virginia and Kentucky south to South Carolina.

Coker Builders, Inc., Box 8, Turbeville, South Carolina 29162. Houses and apartments of 900 sq. ft. to 2,500 sq. ft. are produced for builders and developers according to customers' plans; sold in a 250-mile radius of the company's plant.

Contempri Homes, Stauffer Industrial Park, Taylor, Pennsylvania 18504. Houses up to 1,422 sq. ft. are sold in eastern states including New England, west to West Virginia.

†***Continental Homes,*** Division of Wylain, Inc., P.O. Box 13106, Roanoke, Virginia 24031. Makes panelized as well as modular houses up to 3,000 sq. ft., with retail prices from approximately $25,000 to more than $100,000. Also makes custom house packages according to buyer's plans, commercial structures, and schools. Sells in much of the East and Midwest, south to South Carolina.

†***Continental Homes of New England,*** Division of Wylain, Inc., Daniel Webster Highway, South, Nashua, New Hampshire 03060. Like its sister company above, it makes a variety of both modular and panelized houses, also vacation and multifamily houses; sells in New England and mid-Atlantic states.

DeLuxe Homes of Pa., Inc., 9th and Oak Streets (P.O. Box 323), Berwick, Pennsylvania 18603. Makes single-family, ranch, and raised-ranch houses up to approximately 1,500 sq. ft. Sells up to 300 miles of its plant, or north to New England and west to eastern Ohio.

Dynamic Homes, Inc., Box 875, Detroit Lakes, Minnesota 56501. Its houses are especially attractive and are sold on the basis, "All you do . . . is move in." Houses range approximately from $20,000 to $100,000; also makes multifamily houses; sells in north central United States west to Montana.

†***Havelock Homes Corp.,*** P.O. Box 249, Havelock, North Carolina 28532. Houses up to 1,500 sq. ft.; sold in Maryland, Virginia, North and South Carolina.

†***Haven Homes, Inc.,*** Box 178, Beech Creek, Pennsylvania 16822. Houses up to approximately 1,500 sq. ft.; sold in a 200-mile radius.

Insta Housing, Inc., RR1, Scipio, Indiana 47273. Houses up to 1,600 sq. ft.; sold in a 200-mile radius.

Intermountain Precision-Bilt Homes. See Precision-Bilt Homes, page 160.

Interstate Homes, Inc., 1840 So. 700 West St., Salt Lake City, Utah 84104. Houses up to 3,000 sq. ft.; sold up to 500 miles of Salt Lake City.

Kensett Homes, Kensett, Iowa 50448. Its houses range up to approximately 1,600 sq. ft.; sold up to 250 miles from plant.

†***Key-Loc Modular Homes,*** Box 226, Suncook, New Hampshire 03275. Makes modular houses up to 2,500 sq. ft. with six basic styles (including ranches, split-levels, and Cape Cods) and more than 40 basic models of chiefly traditional design. Models are easily varied to customers' specifications as a result of using modular building units that are interchangeable for design purposes. Also makes "customized" houses, i.e., factory houses designed and built to customer's plans and specifications. Houses are shipped 90 percent complete requiring only foundation and assembly by factory crews. Sells through a builder-dealer network throughout New England and eastern New York. Call 800 555-1212 for its toll-free telephone number (or the toll-free telephone number of anyone else, not just house manufacturers, in the country).

Keyway Homes, Inc., 3584 West Bath Road, Perry, Michigan 48872. One-, 2-, and 3-level houses up to 1,362 sq. ft.; sold only in Michigan.

Kropf Manufacturing Co., P.O. Box 226, Goshen, Indiana 46526. Also makes mobile and custom houses, in other words, from buyers' plans; sizes up to 1,800 sq. ft.; sells east of the Mississippi River. Kropf is noted for high-quality design and construction.

L.C.S. Homes, Inc. (formerly Levitt Homes), Kalama River Road, Fountain Valley, California 92708.

Lancer Homes, Inc., P.O. Box 1200, Corona, California 91720.

Lebanon Homes, Inc., R.D. 2, Route 422, West-North Side, Myerstown, Pennsylvania 17067. This is an "informal shirtsleeve company" that also makes panelized houses, sizes up to 2,400 sq. ft.; sells up to 50 miles of its plant.

†***Manufactured Homes, Inc.,*** 29269 Lexington Park Drive, Elkhart, Indiana 46514. Houses up to 1,400 sq. ft.; sold up to 600 miles from factory.

†***Namretaw Manufacturing Co.,*** Domingo Rt. 60-A, Algodones, New Mexico 87001. One-piece modular houses of 960 sq. ft. to 1,700 sq. ft., selling for less than $24 per sq. ft. in 1979; sold in New Mexico.

Nanco Corp. See under panelized houses, page 168.

†***Nanticoke Homes, Inc.,*** P.O. Box F, Greenwood, Delaware 19950. Houses up to approximately 1,800 sq. ft.; sold within 50 miles of the plant.

†***Nationwide Homes, Inc.,*** P.O. Box 5511, Martinsville, Virginia 24122. Produces panelized as well as modular houses, up to approximately 3,400 sq. ft. Houses are delivered and installed by company crews in mid-Atlantic and southeastern states.

New Castle Homes, Inc., 432 North Memorial Drive, New Castle, Indiana 47362. Also makes panelized houses and commercial mobiles. Sells modulars in Indiana, Illinois and Kentucky, but panelized only within a short distance of New Castle.

Newport Homes, Route 443, Pine Grove, Pennsylvania 17963. Houses up to 1,500 sq. ft.; sold in northeastern states from Virginia north to Maine.

North American Housing Corp., P.O. Box 145, Point of Rocks, Maryland 21777. Offers a good variety of traditional and contemporary styles, 1-, 1½-, and 2-story models, up to 2,600 sq. ft. Sells in the East from New Jersey south to Virginia.

Ozark Homes Manufacturing Co., Inc., Rt. 2, Box 30A, Neosho, Missouri 64850. Makes houses up to 1,900 sq. ft., also makes classrooms, motel, and church units. Ships up to 300 miles of its plant.

Poloron Homes of Pennsylvania, 74 Ridge Road, Middleburg, Pennsylvania 17842. One of the larger manufacturers in the East, Poloron offers close to 150 standard plans (with modifications made on request) ranging from 980 sq. ft. to more than 2,000 sq. ft.; also makes mobile and panelized homes; sells in New England, south to Virginia and west to Ohio.

Porta-Build Inc., 100 Industrial Blvd., Washington, Georgia 30673. Houses up to 1,848 sq. ft.; also makes mobile homes and sells in the southeastern states.

Precision-Bilt Homes, 2525 No. Hiway 89-91, Ogden, Utah 84404. (Also called Intermountain Precision-Bilt Homes.) A large western manufacturer, it offers a variety of plans of different styles up to 1,600 sq. ft., though sizes may be increased in the future. Its construction is noted for earthquake-resistant strength; sells through 160 dealers up to 500 miles from Ogden, Utah.

†***R-Anell Homes, Inc.,*** Denver, North Carolina 28037. See Century Homes above.

Redy-Kwik Homes, Inc., 2914 Clifty Drive, Madison, Indiana 47250. Makes houses up to 1,392 sq. ft.; sells in Midwest up to 300 miles of its plant.

Rochester Homes, Inc., P.O. Box 587, Rochester, Indiana 46975. Makes mobile as well as modular homes up to 907 sq. ft.; sells in eight states in Midwest and East.

Rollohome Corporation, 115 E. Upham St., Marshfield Wisconsin 54449. Makes mobile and modular houses up to 1,456 sq. ft., sells in west north-central states.

Sandler-Bilt Homes, Old Highway East, Boone, Iowa 50036. Houses up to 3,000 sq. ft., in variety of styles and plans, 1-, 1½-, and 2-story models. Makes panelized houses, too; sells in Iowa and surrounding states.

Silvercrest Industries, Inc., 8700 Stanton Avenue, Buena Park, California 90620. One of the larger manufacturers and also makes mobile homes; houses up to 2,100 sq. ft.; sold in eight western states.

Simplex Industries, Inc., 2300 Oram Street, Scranton, Pennsylvania 18504. Makes houses up to 1,900 sq. ft.; sells in the East and New England.

Sterling Corporation, P.O. Box 415, White Pigeon, Michigan 49099. Houses up to 1,440 sq. ft.; sells in the Midwest east to Pennsylvania.

Stratford Homes, Inc., Stratford, Wisconsin 54484. Makes very attractive houses in a variety of types and architectural styles to 2,200 sq. ft.; sells in the upper Midwest west to Montana.

†***Styles Homes, Inc.,*** P.O. Box 9, Scottsville, New York 14546. Has variety of houses up to 1,800 sq. ft., and sells in the East and New England, west to Ohio and West Virginia.

Summey Building Systems, Inc., P.O. Box 497, Dallas, North Carolina 28034. Makes panelized houses, too; traditional plans from 800 sq. ft. to 2,400 sq. ft. Sells in North and South Carolina and Tennessee.

Tidwell Industries, Inc., P.O. Box 679, Haleyville, Alabama 35565. Makes houses up to 1,680 sq. ft., including "energy-saving models"; also makes mobile homes, and sells in 21-state area in the South, East and Midwest.

Torch Industries, Inc., P.O. Box 2268, Elkhart, Indiana 46515. Makes mobile as well as modular, up to 1,700 sq. ft., and says it will sell "almost anywhere."

Travelo Homes Company, 1320 S. Graham Road, P.O. Box 1427, Saginaw, Michigan 48605. Houses of 1,056 sq. ft. to 1,345 sq. ft., sold in the Midwest south to North Carolina and east to Pennsylvania.

†***Unibilt Industries, Inc.,*** P.O. Box 373, 4671 Poplar Creek Road, Vandalia, Ohio 45377. Makes and sells one-family houses up to 1,420 sq. ft., and multifamily housing; sells through builder-dealers in Ohio, Indiana, Kentucky, and West Virginia.

†***Union Manufacturing and Supply Co., Inc.,*** P.O. Box 1696, Fort Collins, Colorado 80522. Makes panelized, as well as modular houses, 850 sq. ft. up to 3,000 sq. ft.; ships up to 500 miles.

Unitized Systems Company, Inc. (USCO Incorporated), P.O. Box 127, Hwy. 58, East, South Hill, Virginia 23970. Houses up to 1,620 sq. ft.; sold in the "eastern seaboard states."

†***Wakefield Homes, Inc.,*** P.O. Box 729, Spring Hope, North Carolina 27882. Houses up to 2,000 sq. ft.; sold in 400-mile radius.

Western Homes Corp., P.O. Box B, North Bend, Nebraska 68649. Makes both mobile and modular homes up to approximately 2,300 sq. ft.; sells in midwestern and north central states.

Panelized Houses

Panelized has to do with the way a house is made in a factory and shipped, as described in chapter 5. It's a house package, made by a large number of manufacturers, in which the walls and sometimes the floor and roof are made in panels, generally 8 feet high and from 4 to 40 feet long. The walls are usually covered on the outside with a skin of sheathing and outside siding (the exterior surface), but may be open or closed on the inside.

The panels are made so they can be quickly connected together at the site and the house shell can be erected, covered, and closed in a day or two. Virtually every size, type, and architectural style of house is available in a panelized package from factories located in every part of the country.

†***Acorn Structures, Inc.***, Box 250, Concord, Massachusetts 01742. Acorn is noted for its handsome houses, including vacation and solar heated models, in varying sizes up to approximately 3,200 sq. ft. Its major suit is top-notch contemporary design. (See photos in chapters 2, 3, and 9.) Its packages generally include all materials for the house shell and assorted additional parts; its builder dealers usually erect the house shell at least, up to a completed house, depending on buyers' preferences. It also makes light commercial and special-purpose structures; sells in eastern seaboard and southern states west to Illinois.

Active Homes Corporation. See under Modular Houses, page 157.

American Standard Homes Corporation, P.O. Box 4908, Martinsville, Virginia 24112. Makes five different lines of houses up to 3,000 sq. ft. in a large recently expanded plant. Offers more than 150 plans; also makes multifamily and vacations houses, and commercial buildings. Ships to 14 mid-Atlantic states up to 500-mile radius of its plant.

American Timber Homes, Inc., P.O. Box 496, Escanaba, Michigan 49829. Makes panelized and precut houses and commercial structures up to 11,000 sq. ft. It emphasizes houses of rustic design (with rough-sawn wall siding, for example), and tough, durable construction with much use of cedar timber because, it says, of its "low-maintenance." Sells east of the Rockies.

Andrus Homes, Inc., P.O. Box 52044, Lafayette, Louisiana 70505. Makes panelized houses for sale in southern Louisiana.

Armstrong Homes, 2715 Auburn Way No., Auburn, Washington 98002. Makes panelized houses of "unlimited" sizes for sale in Washington state.

Barden Homes, Middleport, New York 14105. Houses ranging from small 500 sq. ft. up to large 3,240 sq. ft. models; sells in New England south to Pennsylvania.

Berkshire Construction Company, Inc., Falls Village, Connecticut 06031. This small company is a "custom builder" with more than 950 home plans, from small to large, that are panelized and erected by a company crew; sold in a 65-mile radius of company plant.

Blink Lumber Company, 1379 Comstock, Marne, Michigan 49425. Sells panelized houses up to 150 miles from its plant.

Blue Ridge Homes, P.O. Box 210, Frederick, Maryland 21701. Panelized houses from 980 sq. ft. to 5,500 sq. ft., sold in Pennsylvania, Delaware, Maryland, Virginia, and West Virginia.

†***Boise-Cascade Corporation Housing Division.*** See under Kingsberry Homes, page 167.

Carter Construction & Engineering Co., Inc., P.O. Box 849, Yadkinville, North Carolina 27055. Attractive vacation house packages in hexagonal shapes, up to 1,000 sq. ft.; sold in all the U.S., Canada, and the Caribbean.

†***CBS Homes, Inc.,*** 184 Main St., La Crescent, Minnesota 55947. Panelized houses of "unrestricted" sizes are made and sold in Iowa, Wisconsin, and the Dakotas. Offers a variety of styles, including traditional to contemporary and, in all, more than 3,000 different house plans. It usually erects house and installs all parts except for plumbing and wiring which are provided locally; also makes multifamily and commercial structures.

Chase Barlow Lumber Company, P.O. Box 32038, Louisville, Kentucky 40232. Panelized houses up to approximately 3,000 sq. ft.; sold within 125 miles of Louisville, Kentucky.

Clermont Homes (Clermont Lmbr. Co.), 105 Water St., Milford, Ohio 45150. Specializes in making panelized houses to builders' and home buyers' plans; sells up to 35 miles from Cincinnati, Ohio.

†***Components, Inc.,*** 4400 Homerlee Avenue, East Chicago, Indiana 46312. A large manufacturer of panelized houses, multifamily, and commercial buildings from builders' plans; ships 150 miles from Chicago.

†***Continental Homes, Division of Wylain, Inc.*** See under Modular Houses.

Custom-Made Homes, 416 So. Robinson St., Bloomington, Illinois 61701. Houses up to 6,000 sq. ft. (which is large); sold in Illinois.

Dana Corporation, Box 74, Essex Junction, Virginia 05452. Panelized houses shipped up to 60 miles of plant.

Deck House, Inc., 930 Main Street, Acton, Massachusetts 01720. It is known for houses of striking contemporary looks. (See photo chapter 2.) They're designed especially for low-cost energy use and are characterized by much use of natural wood, outside decks, and large airy interiors. A building system of standard components and post-and-beam construction permit good design flexibility for buyers. House sizes up to 3,000 sq. ft.; sells nationwide.

Derickson Co., Inc., 1100 Linden Ave., Minneapolis, Minnesota 55403. Panelized houses "1,200 sq. ft. and up," shipped and site-assembled by company crews up to 400 miles from Minneapolis.

Duke Millwork, Inc., Thornwood, New York 10594. Makes attractive panelized houses of traditional and contemporary design; ships up to 100 miles.

Dynamic Homes, Inc., Box 875, Detroit Lakes, Minnesota 56501.

Empire Homes, Inc., Division of Janssen/Dakota Corp., 206 South Frontage Road, South Dakota 57004.

Endure Products, Inc., P.O. Box 660666, Miami, Florida 33166. Also makes commercial and industrialized buildings.

E. S. Homes, Highway 30 West, Toledo, Iowa 52342. Houses up to approximately 2,000 sq. ft.; ships in 100-mile radius of plant.

†***Fabricon Corp.,*** 1780 Rohrerstown Road, Lancaster, Pennsylvania 17601. Panelized houses up to 2,500 sq. ft.; ships 250 miles from Lancaster, Pennsylvania.

†***Fleetwood Homes, Inc.,*** Worthington, Minnesota 56187. Panelized houses made chiefly from buyer's plans and specifications; ships in 400-mile radius.

Ivan R. Ford, Inc., McDonough, New York 13801. One of the very oldest firms in the field, Ivan Ford makes closed-panel houses up to approximately 2,000 sq. ft., offers variety of traditional models; ships to New York, New Jersey and Pennsylvania. It ceased house production just prior to publication of this book but is listed because it may start up again.

Gold Star Industries, 32 New Road, Madison, Connecticut 06443.

Green Mountain Homes, Royalton, Vermont 05068. Makes very attractive modern houses, chiefly two-story models, up to 3,000 sq. ft. Offers an excellent low-cost solar heating option that can cut winter fuel bills as much as 60 percent, as described in chapter 13; ships to all states east of the Mississippi.

Habitat. See American Barn/Habitat, page 172.

Haida Hide Inc., 19237 Aurora Ave., N., Seattle, Washington 98133. Makes houses up to approximately 2,200 sq. ft., includes chalet and gambrel designs; ships nationally.

Helikon Design Corp., Cavetown, Maryland 25425. Makes hexagon-shaped houses; in other words practically round, with one or two levels, up to 2,400 sq. ft.; ships up to 400 miles.

Hercon Homes, 105 Main Street, Wayne, Nebraska 68787. Makes traditional houses up to 3,000 sq. ft., also multifamily units; ships up to 150 miles.

Heritage Homes. This is a series of houses made and sold by independently owned manufacturers located in different parts of the United States and Canada. Each uses the same house plans and specifications which are designed by Heritage Homes International, 4850 Boxelder Street, Murray, Utah 84107. The houses are chiefly of traditional design and range in size from approximately 800 sq. ft. to 3,500 sq. ft. Company welcomes capable do-it-yourself buyers who want to assemble or complete any portion of the house package. Materials and factory parts provided vary from a basic package up to everything needed for a completed house, depending on options chosen by each buyer.

Heritage Homes of Thomasville, Inc., 1100 National Highway, Thomasville, North Carolina 27360. Makes Heritage houses, noted above, for sale in southeastern states.

†***Heritage Homes of New England, Inc.,*** P.O. Box 698, Westfield, Massachusetts 01085. Makes Heritage houses, noted above,

and also material packages for builders; sells through some 60 builder-dealers in New England and northeastern New York.

Hexagon Housing Systems, Inc., 905 N. Flood St., Norman, Oklahoma 73069. Houses and apartment units 700 sq. ft. and up; can ship "anywhere."

Homecraft Corporation, Interstate I-85 and U.S. #58, South Hill, Virginia 23970. Makes a wide variety of more than 250 different houses in both traditional (such as southern colonial) and contemporary, up to approximately 3,000 sq. ft. in size, in a highly efficient new plant. It uses open-panel construction; sells through company salesman and some 350 builder-dealers in states from New York south to Florida and west to Ohio.

†***Home Manufacturing & Supply Company,*** 4401 E. 6th Street, Sioux Falls, South Dakota 57103. Houses up to 5,000 sq. ft.; sold in various parts of South Dakota, Minnesota, Iowa, and Nebraska.

†***Intermountain Precision-Bilt Homes, Inc.*** See Precision-Bilt Homes, page 160.

†***Kingsberry Homes, division of Boise-Cascade Corporation,*** 61 Perimeter Park, Atlanta, Georgia 30341. One of the most respected manufacturers in the business, Kingsberry is a large manufacturer that offers a wide line of different kinds of houses, including handsome contemporary models; photos are shown in this book, including construction photos shown in chapter 5. Houses from 700 sq. ft. to 3,300 sq. ft.; has plants in Pennsylvania, Virginia, Iowa, and Oklahoma as well as Georgia; sells through builder-dealers in states west to Utah. Its basic package and house shell usually must be erected by a professional builder, but buyers have the option of completing various parts of the interior.

Loch Homes, Inc., 551 Packerland Dr., Green Bay, Wisconsin 54303. Offers a variety of houses from small to very large, up to 5,000 sq. ft.; both traditional and contemporary designs; sold in Midwest from Michigan to Iowa and Minnesota.

Mid-America Homes, Inc., 2821 East Lincoln Highway, Merrillville, Indiana 46410. Offers line of its own traditional houses and is also a custom manufacturer, i.e., "If we have your own plan, we will build to your specifications . . . any size." Ships up to 200 miles.

Nanco Corporation, P.O. Box 1475, Bellevue, Washington 98009. Makes "panelized-modular" houses; chiefly 3,000 sq. ft.; sells in western states.

†***National Homes Corporation,*** P.O. Box 680, Lafayette, Indiana 47902. One of the largest house manufacturers, it sells a line of mainly traditional houses in nearly every size; has six plants and ships to all states from Atlantic coast to the Rockies.

†***Nationwide Homes, Inc.*** See page 160.

New Castle Homes, Inc. See page 160.

New Century Homes, Inc., P.O. Box 825, Lafayette, Indiana 47902. Traditional houses up to approximately 4,000 sq. ft., with standard floor plans; sold in 250-mile radius of plant.

New England Components. See Techbuilt below.

Neidermeyer-Martin Co., 1727 NE 11th Avenue, Portland, Oregon 97212. This is an old-line company in timber and construction products business that makes both panelized and precut houses of virtually all sizes; it says it ships "worldwide."

Northern Counties Lumber, Inc., P.O. Box 97, Upperville, Virginia 22176. Makes panelized houses of both traditional and contemporary styles, including vacation houses of especially clean design; sells mostly direct to consumers for do-it-yourself or contractor completion; sizes up to approximately 4,000 sq. ft.; ships up to 400 miles.

†***Northern Homes, Inc.,*** 10 laCrosse St., Hudson Falls, New York 12839. Makes particularly attractive houses, including vacation models, in a wide range of traditional and contemporary styles up to 3,200 sq. ft. (see photo chapter 1). Its houses are particularly well-made, and it offers shell packages for do-it-yourself buyers to fairly complete packages, depending on buyers' preferences; plus a number of design and material options. It also makes farm and commercial buildings and "pole structures," all of wood construction; sells through a network of builder-dealers in New England south to Virginia and west to Ohio. Also has a plant on Route 2, Box 15, Chambersburg, Pennsylvania, 17207.

†***Pease Company,*** Hamilton, Ohio 45023. One of the older,

more established home manufacturers, Pease makes precut as well as panelized houses for both vacation and year-round models, and offers very attractive designs (see photo, chapter 3); sizes up to 5,000 sq. ft. "plus." Its house packages are broken down into two parts: "the rough framing" load, delivered first, which contains floor system, exterior wall panels, trim and siding, interior partitions, roof system, windows, exterior doors, exterior cornice, etc.; and the "finish" load, which contains interior trim, doors, flooring, kitchen and bathroom cabinets, vanities, interior hardware, etc. Also makes multifamily units and components for builders; ships up to 400 miles from Cincinnati, Ohio.

F. S. Plummer Co., Inc., 25 Mechanic St., Gorham, Maine 04038. A comparatively small company that makes panelized houses up to 3,000 sq. ft.; ships within a 100-mile radius of its plant in southern Maine.

Precision-Bilt Homes. See page 160.

Richmond Homes, Inc., 1325 Bridge Ave., P.O. Box 336, Richmond, Indiana 47274.

Rondesics Homes Corp., 527 McDowell Street, Asheville, North Carolina 28803. Makes multisided (round) houses with basic building units that can be built as is, or stacked or connected together to provide virtually any size you may want; both vacation and year-round houses. Ships throughout the country and overseas.

St. Marys Precision Homes, Box 597, St. Marys, Pennsylvania 15857. Offers a standard line of traditional houses, and also says, "We specialize in custom building"; in other words, packages made according to customer's plans and specifications of "any size." Ships 75 miles from its plant.

Sandler-Bilt Homes, Old Highway East, Boone, Iowa 50036.

†***Scholz Homes, Inc.,*** 3103 Executive Parkway, Toledo, Ohio 43606. One of the larger manufacturers, Scholz is noted for its impressive luxury houses in the $100,000-and-up price range. It offers traditional designs (including southern Colonial, Georgian, and French chateau) and contemporary models, all with much attention to authentic detail (see photo chapter 1). Many are sold with options, such as spiral staircases and sunken living rooms. Scholz is close to being the Tiffany of manufactured houses. Its prices are not cheap

but still less than the cost of having a comparable custom house designed and built individually for yourself. It also makes custom houses for customers, and multifamily and commercial building units; ships up to 600 miles from its plant and to Colorado and Wyoming.

Shelter-Kit Incorporated, Dept. HB, Franklin Mills, Franklin, New Hampshire 03235. Makes a system of house-building components that provide finished structures ranging in size from one-room cabins up to large houses, depending on the number of units you may want to combine. They are designed for easy do-it-yourself assembly by buyers. Complete packages are sold throughout New England, other packages throughout the country.

Solartran Corp., P.O. Box 496, Escanaba, Michigan 49829. Makes houses up to 3,780 sq. ft.; sells through distributors in the South and other contiguous states.

†***Standard Homes Company***, P.O. Box 1900, Olathe, Kansas 66061. Sells house packages from company plans up to 2,400 sq. ft., also turns out houses from customers' plans; ships to much of the Midwest, south to Arkansas and Oklahoma.

Stimpert Enterprises Inc., 501 Burnside St. S.E., Sleepy Eye, Minnesota 56085. Makes traditional houses of "936 sq. ft. and up"; sells in Wisconsin and Minnesota.

†***Suburban Homes***, P.O. Box 428, Valparaiso, Indiana 46383.

Summey Building Systems, Inc. See page 161.

†***Swift Industries, Inc.***, 1 Chicago Avenue, Elizabeth, Pennsylvania 15037. Makes precut as well as panelized houses (called Lincoln Homes), 750 sq. ft. "and up"; also makes components for others' houses; ships up to 400 miles from Pittsburgh, Pennsylvania.

Techbuilt (sister company of New England Components), 585 State Road, North Dartmouth, Massachusetts 02747. One of the old original "prefab" makers, Techbuilt is famous for its strikingly handsome contemporary houses initiated by top architect Carl Koch. Though comparatively inactive for some years, it is now under new management and makes traditional houses (sold by New England Components), as well as contemporary styles. Houses include small to large models; sold in most parts of the East and New England.

Tri-State Homes, Inc., Highway 51 South, Mercer, Wisconsin 54547. Makes traditional houses of 864 sq. ft. and up; also makes apartment and motel units; sells in Wisconsin and five surrounding states.

†***Union Manufacturing and Supply Co.*** See under Modular Houses.

Universal Homes, Camden, Ohio 45311. Makes traditional houses from 960 sq. ft. to large sizes of traditional design; sold in 500-mile radius of plant.

U. S. Homes, 5390 Second Avenue, Des Moines, Iowa 50313. Makes a wide line of medium- to large-size houses of traditional and contemporary houses, including some very attractive, well-designed models. It is one of the few manufacturers that publishes a list of its home buyers and their addresses in a 24-page booklet; literally thousands are listed. Sells through local builder-dealers in Iowa, Illinois, Missouri, and Nebraska.

†***Wausau Homes, Incorporated,*** P.O. Box 1204, Wausau, Wisconsin 54401. One of the largest manufacturers of regular factory-made houses other than mobile homes, Wausau offers a wide variety of plans, styles, and sizes. Most of its houses are erected and assembled by its own crews; it sells in much of the Midwest and north central states.

Weakley Manufacturing Co., 925 Buckeye Avenue, Newark, Ohio 43055. Makes houses of 1,000 sq. ft. to 3,000 sq. ft.; sells in Ohio.

†***Wick Building Systems, Inc.,*** 6409 Odana Road, Madison, Wisconsin 53719. It offers a variety of houses, called Wick Homes, which are made in various parts of the country and sold through more than 500 builder-dealers in nearly every state. Makes a variety of different sizes, types and styles of houses variously sold in different kinds of packages and with numerous options, including do-it-yourself completion by buyers. Characteristics of the houses vary according to geographical demands; its plants are located in Georgia, Michigan, Missouri, Montana, North Carolina, Texas, and Wisconsin.

†***Yankee Barn Homes, Inc.,*** P.O. Box 1, Grantham, New Hampshire 03753. A big feature of its houses is the use of antique framing timbers, the same yellow pine that years ago went into masts

and bridge trestles, collected from old barns and mills and sawn and notched for re-use. Vaulted ceilings and excellent window design give feeling of bright spaciousness. Savings can be made if owner finishes the interior; sold largely in New England and mid-Atlantic states.

Precut Houses

As the word suggests, a precut house package has much if not all of the lumber for a house shell precut to the necessary size and shape so that it can be easily nailed and assembled to form a finished house. Other parts of the house, such as interior partitions and kitchen and bath cabinets, may or may not be supplied with the package, depending on the manufacturer. In most cases, the heating, plumbing, and wiring must be provided and installed locally by the buyer or his contractor. People who intend to build their own houses, including subcontracting part or all of the job, will find that a precut house most lends itself to saving the most money by using do-it-yourself labor.

American Barn, 123 Elm Street, South Deerfield, Massachusetts 01373.

American Timber Homes. See page 163.

†***Cedar Forest Products Company,*** 107 W. Conden St., Polo, Illinois 61064. Makes precut houses chiefly of western red cedar, up to 2,500 sq. ft.; sells from the eastern seaboard to the Rockies.

Cluster Shed, Inc., P.O. Box 1358, Claremont, New Hampshire 03743. A division of Timberpeg, mentioned below, it makes kits that provide the shell structures for modest vacation cabins; or two or more can be neatly combined for a larger house or other structure; or can provide an additional bedroom, studio or other room that can be easily added to an existing house. Ideal for do-it-yourself interior completion by the buyer; ships east to the Mississippi.

International Homes of Cedars, Inc., P.O. Box 268, Woodinville, Washington 98072. Precut house packages from 400 sq. ft.

(small cabin size) up to 5,000 sq. ft.; sold through dealers nationwide.

Miles Homes, 4500 Lyndale Avenue North, Minneapolis, Minnesota 55412; also has plants in Indiana, Georgia, Pennsylvania, and Texas. Is a fairly large manufacturer that specializes in making shell houses for do-it-yourself buyers; offers largely traditional houses in a variety of sizes and styles and special financing to enable people to buy its shell houses and complete them with little or no cash down payment required; sells in all states west to the Rockies.

Neidermeyer-Martin Co. See page 168.

Pacific Frontier Homes, Inc., P.O. Box 1247, Ft. Bragg, California 95437. Makes precut houses with 900 sq. ft. to 1,700 sq. ft., "modified to any size"; shipped "to any destination."

†***Pease Company.*** See page 168.

Pre-Cut Timber Homes, P.O. Box 97, Woodinville, Washington 98072. Makes precut houses.

President Homes, 4808 No. Lilac Drive, Minneapolis, Minnesota 55429. Makes precut houses of traditional designs from "your floor plan or ours"; virtually any size house. Sells in Midwest and north central states.

Ridge Homes, 501 Office Center Drive, Fort Washington, Pennsylvania 19034. Specializes in what it calls "finish-it-yourself" houses made from its factory packages; offers an assortment of house styles and plans.

Scholz Homes, Inc. See page 169.

Serendipity Shelter Systems, Pier 9, The Embarcadero, San Francisco, California 94111. Makes handsome houses of western contemporary design, including A-frame models, particularly for the "leisure" home market. Houses are sold through factory representatives for do-it-yourself or contractor completion; range up to 2,800 sq. ft.; ships nationwide.

†**Southern Structures, Inc.,** P.O. Box 52005, Lafayette, Louisiana 70505. Makes steel-frame houses in a variety of styles, including A-frame models; provides the preengineered steel shell with no wood framing; interior work by buyer or others. Sells mainly in the South but will ship anywhere in the country.

Swift Industries, Inc., see page 170.

Timberpeg, Box 1358, Claremont, New Hampshire 03743. Makes handsome houses with attractive open-beam ceilings and flexible post-and-beam construction; chiefly two-story New England Saltbox and Barn designs. Factory package basically provides the house shell; interior is completed by local factory dealer-contractor or by the buyer; sizes up to approximately 3,000 sq. ft. Also manufactures Cluster Shed units, described above.

Log Houses

Nearly all log houses from manufacturers are sold as precut house packages, though the size and content of the packages vary from maker to maker. Most makers provide all the precut logs for the outside walls, but from there on you must check on all the additional materials needed to complete the house and whether you or the maker provides them. Sometimes all of the rest must be provided by you.

Another difference among log houses is the kind of wood used; cedar and pine are two of the most common kinds. Many log kits are also sold for do-it-yourself completion by the buyers, but, like building other manufactured houses, this is no picnic. Unlike building another kind of manufactured house, however, schooling in building your own log home is available and is recommended before tackling such work, as noted in chapter 7.

There are some bad apples in the log home field. Some are fly-by-night operators getting into the act for a fast killing, as a result of the rising popularity of log houses. So be careful and cautious before buying. Check on the company you may buy from with your local Better Business Bureau and a few knowledgeable people in the field, such as a good building inspector who's been around.

Companies below marked with a double dagger (‡) are members of the Log Home Council of the National Association of Home Manufacturers.

‡***Air-Lock Log Company, Inc.,*** P.O. Box 2506, Las Vegas, New Mexico 87701. Uses ponderosa pine logs for walls and either

pine or fir for framing members (joists, beams, etc.) and the same hollowed-out "air-lock" log system used by National Log Construction Co., described below; ships throughout the South and Southwest to southern California.

‡***Alta Industries Ltd.,*** Route 30, Halcottsville, New York 12438. Makes log houses with 500 sq. ft. to 2,000 sq. ft.; sells through dealers east of the Mississippi.

‡***Authentic Homes Corp.,*** P.O. Box 1288, Laramie, Wyoming 82070. Offers precut log kits for some 30 different house models, including one-story, split-level, two-story, and chalets for year-round or vacation use; up to approximately 3,000 sq. ft.; sold through dealers in the 48 continental states. Kits contain log walls, floor joists, rafters, gaskets, and spikes. The house shell is ordinarily put up by a company dealer, the rest either subcontracted or completed personally by the buyer, but this is recommended only if you have "adequate construction skills." Also sells log garage kits.

‡***Boyne Falls Log Homes, Inc.,*** U.S. 131, Boyne Falls, Michigan 49713. Houses are made of solid northern white cedar, which turns to a silver hue as it ages, uses sill-and-post construction, and walls are 75-percent preassembled before shipment. House packages contain practically everything required for completion except wiring, plumbing, cabinets and masonry, and are covered by a company 10-year warranty against major defects, structural flaws, and rot or insect damage (which is unlikely because of its cedar logs). House sizes from 900 sq. ft. and up, with custom models made to buyer's plans. Ships throughout the country including Hawaii.

‡***Beaver Log Homes.*** Subsidiary of Chism Industries, 110 N. Cleburn, P.O. Box 1966, Grand Island, Nebraska 68801.

‡***Cedar Homes, Inc.,*** P.O. Box 4109, Bellevue, Washington 98004. Makes precut houses with solid western red cedar logs; sells in seven western states.

‡***Colorado Log Homes,*** 1925 W. Dartmouth, Englewood, Colorado 80110. Offers more than 20 different log house plans, including designs with cathedral ceilings and balconies. Uses 6" x 6" Engleman spruce for the logs, Douglas fir for beams and other structural members. Offers a stripped-down kit for basic house shell, including log walls, and more elaborate packages with parts for the

rest of the house structure. Will modify plans to buyers' wants; ships nationwide.

‡***Green Mountain Cabins, Inc.,*** Box 190, Chester, Vermont 05143.

‡***Justus Solid Cedar Homes,*** P.O. Box 91515, Tacoma, Washington 98491.

‡***Lodge Logs by MacGregor,*** 3200 Gowen Road, Boise, Idaho 83705.

‡***Lumber Enterprises, Inc.,*** Star Route, Box 203, Bozeman, Montana 59715.

National Log Construction Co., P.O. Box 69, Thompson Falls, Montana 59873. Offers more than 50 standard log houses, each of which can be modified by buyers, ranging from 400 sq. ft. to 4,000 sq. ft. Also makes custom log house packages to buyers' plans. Uses the "air-lock" method of hollow-bored logs "for even drying and minimum checking and cracking" and superior insulation, the company claims. Ships to all United States and Canada.

‡***New England Log Homes,*** Box 5056, Hamden, Connecticut 06518. It also has a plant in Lawrenceville, Virginia. Offers a variety of standard log houses up to 4,000 sq. ft., which can be modified if desired and makes precut log house packages to buyer's plans and specifications. Uses hand-peeled pine logs with preservative treatment and an interlock at corners. Ships to all states but Hawaii.

‡***Northeastern Log Homes, Inc.,*** Groton, Vermont 05046. Like many other log house makers, it offers a line of standard log houses, subject to buyer modifications, and will also custom make log houses to buyers' plans and specifications; no limitation on house sizes. Uses 6″ x 8″ preservative-treated white pine logs. Ships to all states and Canada.

‡***Northern Products Log Homes, Inc.,*** Bomarc Road, Bangor, Maine 04401. Offers a variety of houses, many with cathedral ceilings and balconies, from 600 sq. ft. to 3,500 sq. ft. It mainly uses 6- to 8-inch white pine logs for walls, tongue-and-grooved and nailed with 10-inch spikes; plus purloin and truss construction. Offers a basic kit that provides house shell and a standard kit that includes additional parts for the rest of the house. Ships nationally and internationally.

Pan Abode, Inc., 4350 Lake Washington Boulevard North, Rentin, Washington 98055. Makes more than two dozen different models, each sold in precut shell with package; solid western red cedar logs especially for do-it-yourself completion; includes special designs for vacation houses; from 320 sq. ft. to more than 2,000 sq. ft.; sells through dealers nationally.

‡***Real Log Homes: made by Vermont Log Buildings, Inc.,*** Hartland, Vermont 05048; ***Arkansas Log Homes, Inc.,*** Mena, Arkansas 71953; ***Carolina Log Buildings, Inc.,*** Fletcher, North Carolina 28732; ***Traditional Log Homes, Inc.,*** Box 250, State Road, North Carolina 28676; ***Real Log Homes,*** Missoula, Montana 59807. The houses sold by each include a variety of plans from 750 sq. ft. up to 2,900 sq. ft., made in five different plants noted above; uses eastern white pine, lodgepole pine, or southern yellow pine, depending on plant location. Ships to all states and Canada.

‡***The Rustics of Lindbergh Lake, Inc.,*** Seeley Lake, Montana 59868. Noted for log houses that are particularly clean and attractive looking. RLL makes log houses to customers' plans and specifications, as well as offering its own plans; no limit on house size. It uses hand-peeled lodgepole pine logs; sells through dealers in most states west of the Mississippi, including Alaska.

Rustic Log Structures, 14000 Interurban Ave., So., Seattle (Tukwila), Washington 98168.

‡***Traditional Living, Inc.,*** Box 202, Hartland, Vermont 05048.

True Log Building, Box 69, Mason's Bay, Jonesport, Maine 04649.

‡***Vermont Log Buildings, Inc.*** See Real Log Homes, above.

‡***Ward Cabin Company,*** P.O. Box 72, Houlton, Maine 04730. Offers some 30 different log houses and also provides log houses designed and built according to customers' plans and specifications. Uses northern white cedar logs, tongue and grooved to fit, no size limitations on houses made; sells through representatives or direct to customers throughout the country.

‡***Western Valley Log Homes, Inc.,*** Box 254, Victor, Montana 59875.

Dome Houses

Like log homes, most dome houses are made and shipped in precut house packages. While log homes appeal to the nostalgic nineteenth-century yearnings of buyers, dome houses appeal to adventurous buyers who want something new and different, in effect, a twenty-first-century house today. They put their money where their taste lies.

There are clearly more traditionalists than adventurers among Americans today. No more than 3,000 dome houses were sold in 1978 (the latest year for which figures were available), according to Bob Casey, president of the National Association of Dome Manufacturers. By comparison, 30,000 log houses were sold in 1978. But if you want to be the first one on your block with a really new and different house, here are dome house manufacturers who can provide one.

Each sells a dome shell kit but not all permit do-it-yourself assembly, since it's not as easy as it may appear. Several makers strongly "prefer" to provide supervision of the foundation, to begin with. Like the foundation for other factory houses, it must be carefully and properly built or else you're in serious trouble. Next step, erection of the geodesic shell, also requires skill and experience, and at least one maker (Domes America) insists that its dome shells be erected by its own people. Its president says that even a supposedly skilled and capable buyer can easily botch it. Putting on a good leak-proof roof is a third critical stage of the construction where mistakes are commonly made and indoor rain will quickly let you know about them. Most makers willingly let buyers complete the rest of their dome houses, and in some cases this is necessary because the manufacturer provides nothing more than the dome shell.

Dome houses are still not big business and the companies in it are, as a result, comparatively small and also going through inevitable growing pains. This makes it particularly important to check carefully on the one you may buy from, be sure of what it furnishes, what you must provide, as well as being sure that the company is dependable. Most of them will ship their domes long distances, but it is best to buy from one who is nearby; this cannot be overstressed.

And it's good, if not essential, to buy only from one that has satisfied dome owners living in their domes nearby where you can see and talk to such buyers *before* you go ahead. If not, you buy a dome at great peril.

Allard Engineering, South Lee, Massachusetts 02160.

Big Outdoors People, Inc., 2201 N.E. Kennedy Street, Minneapolis, Minnesota 55413.

Carter Construction Engineering Company, Inc., P.O. Box 849, Yadkinville, North Carolina 27055.

Cathedralite Domes, P.O. Box 880, Aptos, California 95003.

Domes and Homes, Inc., P.O. Box 365, Brielle, New Jersey 08730.

Domes America, 6 South 771 Western Avenue, Clarendon Hills, Illinois 60514.

Dome West, 181 Pier Avenue, Santa Monica, California 90405.

Domiciles, Inc., Route 1, Numa, Iowa 52575.

Dyna-Dome, 22226 N. 23rd Avenue, Phoenix, Arizona 85027.

Envirotecture, 134 N. Ojai St., Santa Paula, California 93060.

Geodesic Domes, Inc., 10290 Davison Road, Davison, Michigan 48423.

Geodesic Homes, Box 1675, Bailey, Colorado 80421.

Geodesic Structures; see Domes and Homes, Inc., above.

Monterey Domes, Inc., 3777 Placentia, Box 5621-AA, Riverside, California 92514.

Polydome, Inc., 1238 Broadway, El Cajon, California 92021.

Space Structures International, 325 Duffy Avenue, Hicksville, N.Y. 11801.

Synapse, Inc., Box 554, Lander, Wyoming 82520.

Zomeworks, 1212 Edith Boulevard, Albuquerque, New Mexico 87103.

A-Frame House Manufacturers

Unlike log and dome houses, A-frames are not exclusively made by individual manufacturers. They are made as one of a number of different kinds and styles of houses. They are made by, among others, makers of log homes (thus A-frame log houses) and by precut house manufacturers (thus delivered in precut house packages). Because the A-frame was born largely as a vacation house, it is offered today chiefly by such companies, though an increasing number of A-frame houses are made for year-around living.

Bell Aire Log Cabin Manufacturing Co., P.O. Box 322, Bell Aire, Michigan 49615.

Carroll Homes, Inc., 2434 Forsyth Road, Orlando, Florida 32807.

Cedar Chalet Div., Leisuretime Products. See page 157.

Easy A, Division Southern Structures, Inc., P.O. Box 52005, Lafayette, Louisiana 70505.

Forest Houses, Inc., P.O. Box 696, Mesa, Arizona 85201.

Herculean Homes Corp., P.O. Box 357, Benttendorf, Iowa 52722.

Pacific Panel Houses, 7951 2nd Avenue South, Seattle, Washington 98108.

Serendipity. See page 173 (Pre-Cut Houses).

Southern Structures, Inc. See page 173.

TimberLodge, Inc., 1309 Swift, N. Kansas City, Missouri 64116.

Vacationland Homes, Inc., P.O. Box 292, Bell Aire, Michigan 49615.

Mobile Homes

Visiting and seeing actual mobile homes at dealer show rooms or elsewhere, particularly the model you may order, is virtually essential before you buy one. It's the only way you can get a good idea of what you will and will not like. Besides, you may like a particular model seen in a manufacturer's brochure, but it may be available only from one of his distant plants and therefore unavailable in your area. And like cars, some manufacturers also change their models every year.

Thus, a main purpose of the following directory is to tell which makers have factories that sell in your area. That generally means a location no more than 400 miles from you, depending on the manufacturer. Then you write directly to the manufacturer for his up-to-date catalog and prices, his maximum shipping distance, and names of his dealers near you.

An asterisk (*) means that the manufacturer is a member of the Manufactured Housing Institute, 1745 Jefferson Davis Highway, Arlington, Virginia 22202, the national trade association for mobile-home makers.

Two asterisks (**) mean a member of the Western Manufactured Housing Institute, 3855 E. LaPalma Ave., Anaheim, California 92807, the mobile-home association of western producers.

Three asterisks (***) indicate membership in both associations.

The names and addresses of mobile-home manufacturers in Canada can be obtained from the Canadian Mobile Homes Institute, Suite 705, 151 Slater Street, Ottawa, Canada K1P5H3.

Admiration Homes, Inc., P.O. Box 39, Elkhart, Indiana 46515.

*_**All American Homes, Inc.**_, P.O. Box 451, 309 South 13th Street, Decatur, Indiana 46733.

*_**APECO Corporation, Homebuilders Group**_, 2100 Dempster Street, Evanston, Illinois 60204.

Bendix Home Systems**, 61 Perimeter Park, Atlanta, Georgia 30341. Has a dozen plants in different parts of the United States and in Ottawa, Canada.

*_**Brigadier Industries Corporation**_, P.O. Box 567, Sylvester, Georgia 31791.

Burlington Homes, Inc., State Highway 621, St. Clair, Pennsylvania 17970.

*_**Carolina International, Inc.**_, 1 Hurstbourne Park, Suite 703, Louisville, Kentucky 40222.

Catalina Homes, Inc., Douglas, Georgia 31533.

Cavco Industries, Inc., P.O. Box 6556, Phoenix, Arizona 85005.

Century Housing Corporation, P.O. Box 737, Fort Morgan, Colorado 80701.

*****Champion Home Builders Company,*** 5573 East North Street, Dryden, Michigan 48428. Makes mobile homes in 25 manufacturing plants in various parts of the country; sells nationally.

****Chief Industries, Inc.,*** Housing Division, P.O. Box 349, Aurora, Nebraska 68818.

*****The Commodore Corp.,*** Box 300, Danville, Virginia 24541. Has four other plants elsewhere.

****Conner Homes Corporation,*** P.O. Box 520, Newport, North Carolina 28570.

Craftmade Homes, P.O. Box 1185, Henderson, Texas 75652.

Custom Craft Mobile Homes, Inc., P.O. Box 115, Highway 67 South, Clinton, Louisiana 70722.

De Rose Industries, Inc., 4002 Meadows Drive, Suite 116, Indianapolis, Indiana 46205.

****Diversified Manufactured Housing Company (also DMH Company),*** 1517 Virginia St., St. Louis, Michigan 48880.

*****Dualwide Homes,*** 601 East Wooley Road, Oxnard, California 93030.

Elliott Mobile Homes Manufacturing Co., P.O. Box 305, Waurika, Oklahoma 73573.

*****Far West Homes,*** P.O. Box 1238, Yuba City, California 95991.

*****Fleetwood Enterprises, Inc. Housing Group,*** 3125 Myers St., Riverside, California 92523. Has factories in a number of states and sells throughout the United States and Canada.

*****Fuqua Homes, Inc.,*** 7100 South Cooper, Arlington, Texas 76017. Has nine factories in various parts of the country.

*****Guerdon Industries, Inc.,*** P.O. Box 35290, Louisville, Kentucky 40232. One of the largest makers, it has some 20 factories that turn out mobile homes in virtually every part of the country.

Herrli by Garden Place Homes, Inc., Box 848, Elkhart, Indiana 46515.

Homes of Merit, Inc., P.O. Box 1606, Bartow, Florida 33830.

Horton Homes, Inc., P.O. Box 581, Eatonton, Georgia 31024.

Indies House, Inc., P.O. Box 129, Hackleburg, Alabama 35564.

****Kaufman and Broad Home Systems, Inc.,*** 10801 National Boulevard, Los Angeles, California 90064.

*****Kit Manufacturing Company,*** P.O. Box 250, Caldwell, Indiana 83605; P.O. Box 738, McPherson, Kansas 67460; P.O. Box 365, Forest Grove, Oregon 97116.

****Kropf Manufacturing Co.*** See page 159.

*****Lancer Homes, Inc.,*** P.O. Box 1200, Corona, California 91720.

****Liberty Homes, Inc.,*** P.O. Box 35, Goshen, Indiana 46526.

Majestic Industries, Inc., P.O. Box 5577, Texarkana, Texas 75501.

****Mansion Homes Corporation,*** P.O. Box 756, Robbins, North Carolina 27325.

******Marlette Homes, Inc.,*** 3270 Wilson, Marlette, Michigan 48453.

****Mascot Homes, Inc.,*** P.O. Box 127, Gramling, South Carolina 29348.

Nobility Homes, Box 1652, Ocala, Florida 32670.

North East Housing, Inc., Route 26, Oxford, Maine 04270.

Nuway Mobile Home Manufacturing, Inc., P.O. Box 4463, Fort Worth, Texas 76106.

****Parkwood Homes, Inc.,*** 19224 Cty. Road #8, Bristol, Indiana 46507.

****Patriot Homes, Inc.,*** 57420 Cty. Road 3S., Elkhart, Indiana 46514.

Poloron Homes of Pennsylvania. See page 160.

Porta-Build, Inc., 100 Industrial Blvd., Washington, Georgia 30673.

****R-Anell Homes, Inc.,*** Denver, North Carolina 28037. Also see page 161.

******Redman Homes, Inc.,*** 2550 Walnut Hill Lane, Dallas, Texas 75229. Another large manufacturer, Redman had, at last count, 17 different plants in various parts of the country except the Northeast.

Regal Homes, 999 Budd Road, Montevideo, Minnesota 56265.

****Rochester Homes, Inc.,*** P.O. Box 587, Rochester, Indiana 46975.

****Rollohome Corporation,*** 115 E. Upham St., Marshfield, Wisconsin 54449.

****Schult Homes Corp.,*** P.O. Box 151, Middlebury, Indiana 46540; also has plants in Arizona, Florida, Kansas, Kentucky, Maryland, Minnesota, North Carolina, and Texas.

****Shelter Resources Corporation,*** 24200 Chagrin Boulevard, Beachwood, Ohio 44122. Has a number of other plants.

*****Silvercrest Industries, Inc.,*** 8700 Stanton Ave., Buena Park, California 90620.

****Skyline Corporation,*** 2520 By-Pass Road, Elkhart, Indiana 46514. One of the very largest producers, it makes mobile homes in 37 different plants in the country.

Star Homes, Inc., P.O. Box 98, Starr, South Carolina 29684.

*****Sturgis Mobile Homes, Inc.,*** 17061 Compton Ave., Corona, California 91720.

Suncoaster Mobile Homes, 9190 Ulmerton Road, Largo, Florida 33541.

****Tidwell Industries, Inc.,*** P.O. Box 679, Haleyville, Alabama 35565.

****Torch Industries, Inc.,*** 28384 County Road 20 West (P.O. Box 2268), Elkhart, Indiana 46515.

Tropicana Mobile Homes, Inc. (Tropicaire Name Plate), 3310 U.S. 19 N., Clearwater, Florida 33515.

****Victorian Homes, Inc.,*** P.O. #707, Middlebury, Indiana 46540.

Vintage Homes, Inc., 3825 NE Expressway, Atlanta, Georgia 30340.

****Western Homes Corporation,*** 219 E. 6th St., North Bend, Nebraska 68649.

Wick Building Systems, Inc. See page 171.

Winston Industries, Inc., P.O. Box 347, Double Springs, Alabama 35553.

Young American Homes, Inc., P.O. Box 1325, Wingate, Pennsylvania 16880.